# CHEMISTRY & PHYSICS

## Elements and Forces of the World

TESLA COIL

Mr. Adolfo Ayala is an Assistant Professor of Chemistry in the department of Biology and Biochemistry at Belmont Abbey College. He has been teaching high school science and math in some capacity for the past six years. He is a gifted teacher and makes learning math and science understandable and enjoyable. He has also taught courses in Chemistry and Engineering at various small colleges throughout the country. Mr. Ayala earned his MS in Chemical Engineering at the University of Notre Dame after completing his BS in Chemical Engineering at the University of Arkansas. He has a hobby of collecting math instruction books, both old and new, so he has a familiarity with a variety of teaching methods and believes that the older methods provide a better foundation for proficiency in the mathematical arts. Mr. Ayala assists his wife in homeschooling their four children. When he has free time, Mr. Ayala enjoys playing tennis, cooking, gardening, and practicing his woodworking skills. You can often find him reading a book on philosophy or theology and drinking a cup of his freshly roasted coffee.

CHEMISTRY EXPERIMENT IN A LABORATORY

# CHEMISTRY & PHYSICS

## Elements and Forces of the World

Adolfo Ayala

TAN Books
Gastonia, North Carolina

Cover & interior design and typesetting by www.davidferrisdesign.com

ISBN: 978-1-5051-2618-1

Published in the United States by
TAN Books
PO Box 269
Gastonia, NC 28053

www.TANBooks.com

Printed in the United States of America

"*Where were you when I laid the foundation of the earth?*"

—Job 38:4

HOLMIUM NITRATE UNDER THE MICROSCOPE. HOLMIUM IS A RARE EARTH ELEMENT
USED FOR DIFFERENT APPLICATIONS IN ELECTRONICS, LASERS AND GLASS COLORING.

# CONTENTS

# PREFACE

When I think about the scientific study of the natural world, two phrases from the writings of Pope St. John Paul II come to mind:

(1) a rigorous pursuit of truth and

(2) a love of learning.

The first—a rigorous pursuit of truth—describes science and its processes. Scientists make careful observations, design experiments, and collect data to learn more about how the world works. Too often, though, science may seem like something you do in a big research facility with a lab coat.

But we are all scientists!

Anyone can study the living world in a scientific way. From an early age, everyone has a curiosity to understand the world. Think of a baby repeatedly dropping something onto the floor; he is discovering how gravity works! It is this basic curiosity that drives science.

The second piece—a love of learning—also describes what science should inspire. Sometimes science is depicted as a dry, boring set of facts, but nothing could be further from the truth. The world is a fascinating place. I have been interested in the natural world my whole life. This love of nature led me to obtain undergraduate and postgraduate degrees that have allowed me to teach biology classes every day for a living, and yet I am still constantly amazed by the wonders of our world.

There is always something new to learn within the natural and physical sciences. This unit discusses the sciences of chemistry and physics. As a biologist, I know that these concepts have a profound role in understanding biological organisms and systems. There are so many fascinating things to learn about these sciences, and how they affect the world around us.

For example, did you know:

- the reason it is more difficult to walk through water than on dry land is because of a force known as hydrodynamic drag?

- we can see lightning before we hear the clap of thunder because light moves at an astonishing 300,000,000 $^m/s$, while sound only travels at 340 $^m/s$?

- echoes work better when they bounce off a hard surface instead of a soft one?

What an amazing world we live in! This text, written by friend and colleague, Adolfo Ayala, will introduce you to chemistry and physics and some of the rules and laws written into the physical world around us.

*"[Science and faith] each can draw the other into a wider world, a world in which both can flourish."*

—Pope St. John Paul II in *Physics, Philosophy and Theology*

Finally, it is too often assumed in our society today that faith and science act in opposition to one another, that somehow if we learn enough about the world, it would disprove the existence of God. But it is important for each of us to be confident in our Faith and the fact that truth cannot be in opposition with itself.

We read in the *Catechism of the Catholic Church*: "Methodical research in all branches of knowledge, provided it is carried out in a truly scientific manner and does not override moral laws, can never conflict with the faith, because the things of the world and the things of faith derive from the same God. The humble and persevering investigator of the secrets of nature is being led, as it were, by the hand of God in spite of himself, for it is God, the conserver of all things, who made them what they are" (*CCC* 159).

Holy Mother Church teaches us that we can pursue scientific knowledge unafraid. It is my hope that *The Foundations of Science* series will not simply give your family some facts about the world but also instill a curiosity and love of learning in you that you can apply across all the disciplines of your life, both scientific and otherwise.

**Timothy Polnaszek, PhD**

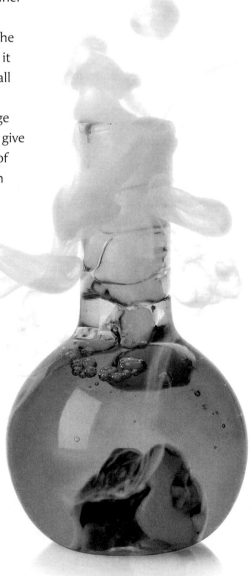

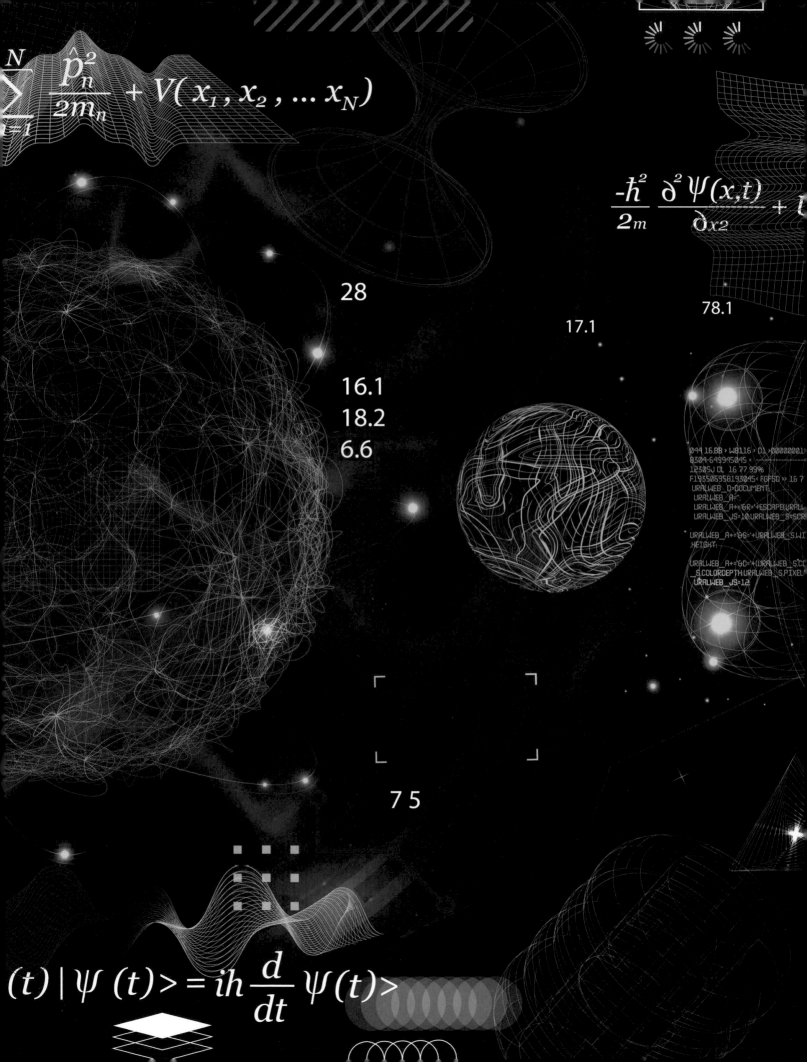

# INTRODUCTION

Most of you have probably heard of the scientific studies of chemistry and physics. But do you really know what they are? Do you know what chemists and physicists study? Well, you are about to find out, because you are about to become junior chemists and physicists!

Chemistry and physics are natural sciences in that they study characteristics of the world. In a sense, we can say they study everything around us. For example, as you got ready to sit down to read this, you probably turned on a light so you could see and shut your bedroom door to block out the sound in your house so you could concentrate. You may have taken a deep breath before cracking open the front cover or brought a snack with you to eat to give you some energy. Believe it or not, with all these actions, you were mixed up in the world of chemistry and physics!

Though these two sciences surround us every day, they are somewhat different from the other topics you have studied in *The Foundations of Science* series because much of what takes place in chemistry and physics happens "below the surface." We don't consciously notice or think about all the ways chemistry and physics are present in our lives like we do when we are studying animals or plants. And yet, every second of our day depends upon the objects of these two natural sciences.

At their core, these sciences are the study of different aspects of *matter* and *energy* (those are two words you are going to hear a lot over the next twelve chapters). Chemistry is the study of the structure of different types of matter, the transformation of one kind of substance into another kind of substance, and the energy changes occurring in these transformations. Physics, meanwhile, studies the way in which matter can be moved in space and the energy changes involved in those movements. This includes the study of sound, light, electricity, and other areas of life that will be familiar to you, only you will understand them on a deeper level when our journey concludes. So without any further delay, let's dive into this fascinating world!

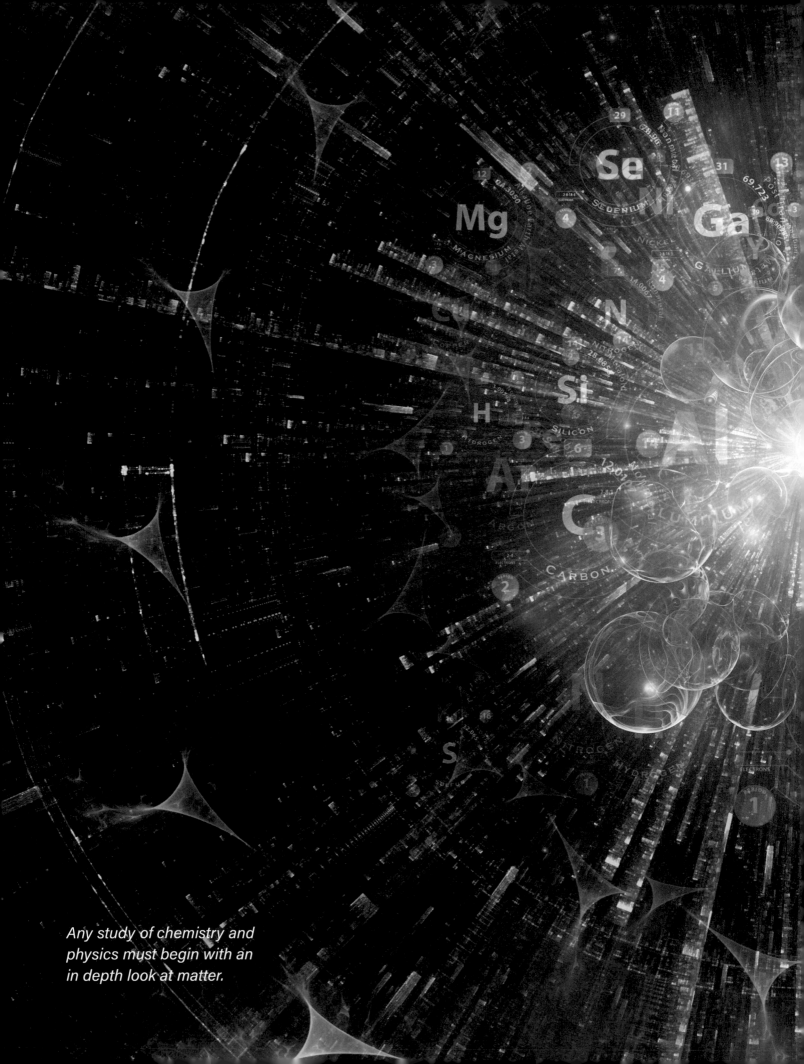

Any study of chemistry and physics must begin with an in depth look at matter.

# CHAPTER

## 1

*WHAT IS MATTER?*

## UNDERSTANDING THE BASICS

A good place to start our study of chemistry and physics would be by defining some of the terms that we use in these branches of science. An interesting thing you will notice as we go through this book is that many of the terms that are used are taken from everyday language, but they are given special scientific definitions. This was done a long time ago as early physical scientists took everyday words and gave them a technical definition when they used them in their work.

Chemistry and physics are the study of matter. With that definition, we introduced two words—*study* and *matter*—that may be familiar, but when talking about these sciences, we won't be using them quite like you are used to. Let's look at each of these terms in turn, starting with matter.

Everything in the material world is made up of matter. In the simplest terms, **matter** is anything that takes up space and can be weighed, which means matter has volume and mass, or density. We've introduced even more terms here so we will need to define those too.

**Volume** refers to the amount of space something occupies (you can think of it like the size of something, as in your dad's volume is more than yours), while **mass** is the quantity of matter (how much matter) in an object. **Density** meanwhile describes *a relationship between volume and mass*—it is the amount of mass packed into a given unit of volume. Think about it this way: density can be used as a descriptor of how "heavy" an object is. A dense object will feel heavy for the amount of space it occupies. Consider a balloon filled with air and a bowling ball. Even though they are approximately the same size

*Physics fun fact:*
*A bowling ball and a balloon can have approximately the same volume but have wildly different densities.*

(the same volume), we know the bowling ball is much heavier (has more mass packed into that volume). Therefore, the bowling ball is the denser object. In another book of this series, we used a different analogy to understand density, comparing two suitcases of the same size: one with three T-shirts packed in it, and another with thirty T-shirts. Though the suitcases have the same volume (they are the same size), the one with thirty T-shirts has a higher density (and therefore weighs more) because there is more mass (more T-shirts) packed into that given volume.

Another way to define mass is by saying it is the volume, or space, of the given matter multiplied by the density of the object (this is Isaac Newton's definition—we will talk more about him in chapter 4). Notice that we have now added a mathematical operation into the definition. This mathematical understanding is one of the defining characteristics of the study of matter in chemistry and physics, but we will come back to this in the next chapter.

Before moving on, let's review what we have learned so far:

- **Matter**: Anything that takes up space (volume) and can be weighed (density/mass).

- **Volume**: The space an object occupies or takes up.

- **Mass**: The quantity (amount) of matter in an object.

- **Density**: The relationship between volume and mass; the amount of mass packed into a given unit of volume; a descriptor of how "heavy" an object is.

## SENSING MATTER AND ITS PROPERTIES

Another way to think about material things, or matter, is that they can be detected using our senses, meaning we can hear, taste, see, smell, or feel matter. However, this does not mean that we can use *all* five senses *all* the time. Some things you can smell but you cannot see, or we can feel air when it is blowing, but our eyes do not see it. Still, matter is discernable by our senses, even if not by all of them. We use our senses to describe objects and to share the experiences with others. We can distinguish a heavier object from a lighter object by holding them at the same time. We can see who runs faster in a race by using our sight to see who crosses the finish line first. Sometimes we use machines to help our senses, as when we use a microscope to see very small things or a telescope to see things that are very far away.

Experience of the natural world through our senses gives us an expectation of how objects will normally behave. If you drop a rock, or a rabbit, or a bucket of water from a bridge into a pond, you know that all three will fall into the water because of gravity. The rock will sink because it is heavy, the rabbit will get wet, and the water in the bucket will mix with the water of the pond. Speaking of water, we know that it can turn into ice if it gets very cold, or steam if it gets very hot. You can differentiate plain water from a drink like tea by looking at it, because the color is different. You can even tell which is which while blindfolded because they taste so different.

Different substances have different tastes, textures, and colors. They also have different densities, different

### Beyond Matter

This is a book about matter, its structure and its changes. But as Catholics, we know that not all that exists is *material* (meaning physical). Yes, we know that all matter, including ourselves, is created by God, yet we humans are not simply material beings. God created us in His image and likeness (see Gn 1:26-27)—this means we are more than just physical creatures. The faculty we have to take the information from our senses and consider higher thoughts with it points to our intellect and to our soul being spiritual and immortal. It is the spiritual nature of our soul that allows us to think beyond the particular material objects present to us. It allows us to abstract from the many tables we have seen what a table is, and allows us to judge what makes a table "good." It allows us to define what matter is and gives us the faculty to judge the different types of matter and categorize it. Cardinal Joseph Ratzinger (later Pope Benedict XVI) wrote that the scientist is not creating anything in carrying out scientific studies but "rethinking" that which has been thought before. In this sense, the scientist is trying to go beyond matter to understand the mind of God.

temperatures at which they freeze or boil, and they differ in how easily they get hot or how quickly they cool off. We call all these descriptors of a substance the **properties** of that given substance; they are like different characteristics, just like you might have certain physical characteristics (tall or short, brown or blond hair, etc.). Everything you have studied in the *Foundations* series is made of matter, from the largest star to the smallest bacteria, from a blue whale to a grain of sand. Each of these things has certain properties.

*The properties of a given substance are like its characteristics. One characteristic of water is that it can be found in three separate states of matter (liquid, solid, gas) depending on what its temperature is.*

## BREAKING DOWN MATTER

If everything in the physical world is made of matter, then could we say that chemistry and physics are the sciences of *everything*? In a way, yes! Some scientists would say that it is possible to treat everything as basic physics and/or chemistry. But, of course, we must further break down things in order to truly study them.

When we look at all the things made of matter, we can make distinctions by characterizing them as either *living* or *nonliving*. Things that have life (animals, plants) are the subject matter of **biological sciences**. Chemistry and physics, meanwhile, look at the material world from the perspective of nonliving matter. This does not mean that chemistry has nothing to say about what happens to the matter in the cells of an organism, but still, we cannot treat the organism as simply a series of chemical reactions. Even when looking at a living organism, chemistry and physics treat the matter in the organism in the same way they treat that matter in a laboratory.

After establishing the limits of chemistry and physics to nonliving matter, we can ask the question: Is all nonliving matter the same? As you might expect, the answer is no. Scientists characterize matter into either *mixtures* or *pure substances*. Let's look at mixtures first.

A **mixture** is something that is made from at least two or more pure substances. There are two different kinds of mixtures: heterogeneous mixtures and homogeneous mixtures. A **heterogeneous mixture** is one in which we can observe the different components of the mixture. Think of a tossed salad with lettuce, tomatoes, cheese, and nuts. The salad is a mixture of all those components, and you can easily see the individual components. Another example would be chocolate chip cookie dough. You clearly see the chocolate chips scattered throughout the buttery dough. But not all heterogeneous mixtures are this easy to identify. Sometimes we need a microscope to see the different components of a mixture. The milk people buy at the store looks like it is all the same throughout, but if you were to put a drop of it under the microscope, you would see little bubbles of fat floating in the watery whey of the milk.

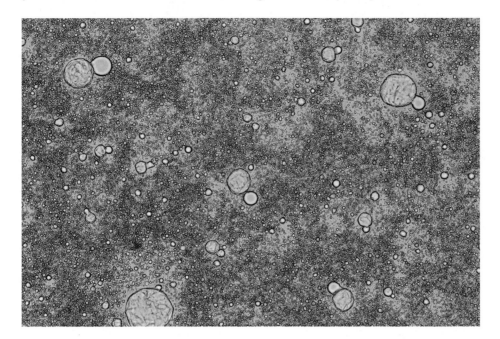

*Under a microscope milk doesn't look quite as delicious! Milk is an example of a heterogenous mixture because we can observe different components within the mixture.*

A **homogeneous mixture** is one in which the different substances that are mixed together *cannot* be observed separately. This mixture may look like a pure substance but is not. The vanilla extract in your pantry is an example of a homogeneous solution with a lot of different substances present. The extract will have water, vanillin, maybe alcohol, maybe sugar, and other possible substances in it. When you pour it into a teaspoon, you do not see the different components, as they are all dissolved by the water in the extract (it just looks like one substance despite being a combination of many).

Both homogeneous and heterogeneous mixtures can be separated into their pure substances (the "whole" can be divided into "parts") using techniques called physical changes. A **physical change** is one that changes the given substance but does not change *the kind of* substance you have. For example, if you had a glass of sugar-water, we can get the sugar out of the mixture by boiling off all the water. Though we have changed the sugar-water, the evaporated water is still water (now in a gas form) and the sugar that remains is still sugar.

Let's now pivot to talking about the other kind of matter: pure substances. **Pure substances** are those that cannot be separated from other substances using physical changes. A pure substance can be changed into other pure substances through what scientists call chemical changes, or chemical reactions (we will discuss these more in a later chapter). There are two types of pure

# WHAT IS MATTER?

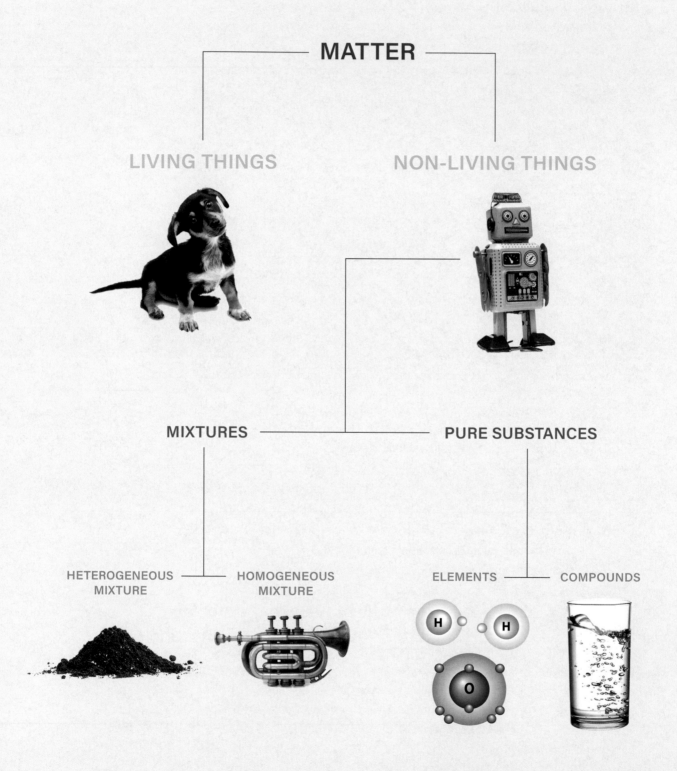

MATTER

LIVING THINGS

NON-LIVING THINGS

MIXTURES

PURE SUBSTANCES

HETEROGENEOUS MIXTURE

HOMOGENEOUS MIXTURE

ELEMENTS

COMPOUNDS

H  H

O

substances: elements and compounds. A **compound** is a chemical substance made up of multiple elements, which can be broken up or joined through a chemical reaction, while an **element** is a substance that cannot be broken into any other type of substance by using chemical reactions. For example, we can take hydrogen gas and combine it with oxygen gas to make water ($H_2O$). Hydrogen and oxygen are elements, and water is a compound. Water can also be broken back into oxygen and hydrogen. The fact that we can break water into elements means it is a compound. We will add a layer to this understanding when we discuss atoms later in the book.

At the beginning of this chapter, we said that chemistry and physics are the study of matter. Now that you have a general understanding of the term matter, we can discuss the method scientists have used and continue to use to study the material world.

*Physics fun fact:*
*Everything in the world is made of matter. Everything!*

## FOUNDATIONS REVIEW

✓ Chemistry and physics are the study of matter. Matter is anything that takes up space and can be weighed, which means matter has volume and mass, or density. Volume refers to the amount of space something occupies, while mass is the quantity of matter (how much matter) in an object. Density, meanwhile, describes a relationship between volume and mass—it is the amount of mass packed into a given unit of volume. Density can be used as a descriptor of how "heavy" an object is. A dense object will feel heavy for the amount of space it occupies.

✓ We can detect matter and understand it (understand how it will behave) using our senses. We do this by observing its properties, which describe its characteristics, things like how it tastes, what it feels like, or what color it is. Different types of matter can also have different densities, different temperatures at which they freeze or boil, and they can differ in how easily they get hot or how quickly they cool off.

✓ Matter can be broken down into two categories: living and nonliving things. Chemistry and physics, for the most part, study nonliving things. Nonliving things can be further broken down into pure substances (elements and compounds) and mixtures (heterogeneous and homogeneous). A mixture is something that is made from at least two or more pure substances.

# Understanding Matter: Form and Accidents

When you see a marble statue, is the fact that the statue is made from marble the only thing that is important about it? Is that what makes it a statue? Or is the shape it has been formed into more important?

Usually, you determine it is a statue because it has been shaped by a craftsman. You can see the features of it and his intention, maybe even the tool marks that show his agency. We can say that the lump of matter (the marble), through the hands of the craftsman, has received a *form*. The lump of matter in the hands of a well-trained artist can be transformed into a beautiful work of art by the agency of the artist (by agency we mean through the free action and intention of the artist—his ability to act for himself and make decisions). We can say that the lump of matter had the potential to be actualized into a statue by the artist who had the ability to do so.

A full understanding of an object must take into account that it is made of the union of matter and form. The matter is what it is made of (marble), and the form is that which makes it to be what it is (a statue). The bench I sit on as I write this is made of wood, but it may just as well have been made of plastic, stone, or metal. In each case, no matter what it was made from, because it has the form of a bench, you would be able to recognize it as a bench.

Similarly, when you look at a dog, you know it is a dog because its matter is "united" to the form of dog. The adjectives that describe a dog (big, small, white, black, etc.), its "properties", in a theological sense, are called the *contingent accidents*. What makes a Labrador Retriever different from a Saint Bernard? The differences between the dogs come from their differences in size, color, musculature, etc.—they have different contingent accidents. They are both clearly dogs, as they have the characteristics of a dog. We would say these characteristics are *essential accidents*, those things that if changed would make it no longer a dog.

If someone asks you what a table is, you would use the *essential accidents* to describe a general table. If you were asked to describe your dining table, you would then give all the *contingent accidents* that are particular to your table. For example, my dining table is made of oak, has a light finish, and has a burn mark right in the middle where my wife's roommate burned it before we got married! In a sense, the essential accidents are general qualities about a thing,

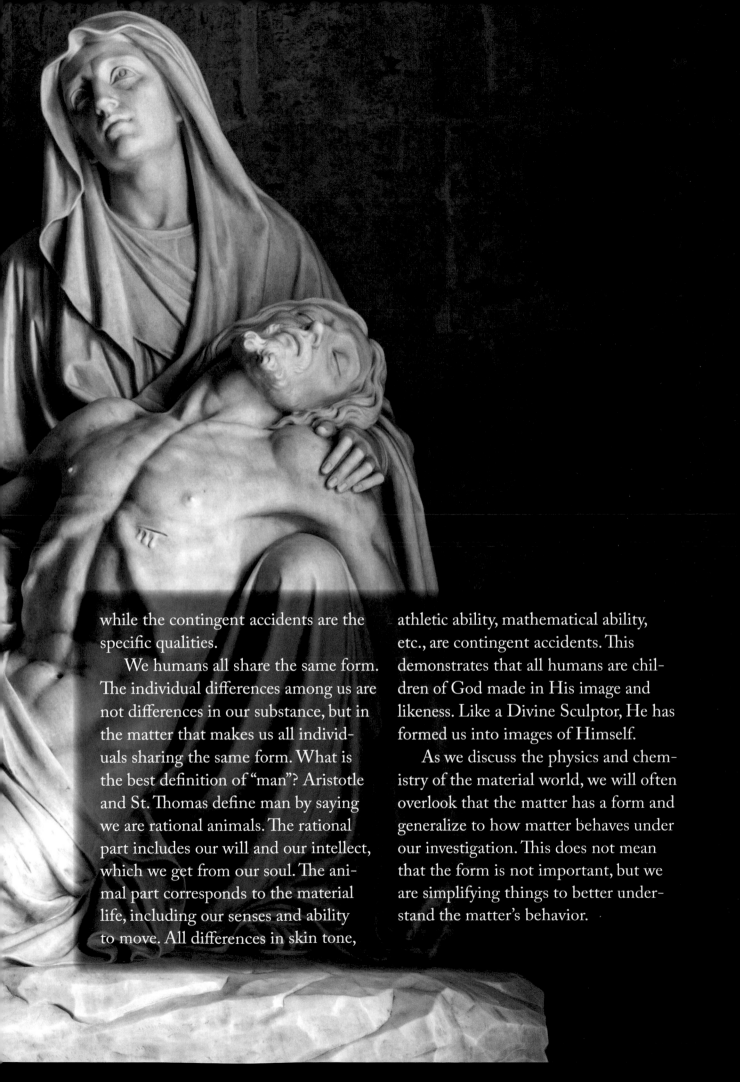

while the contingent accidents are the specific qualities.

We humans all share the same form. The individual differences among us are not differences in our substance, but in the matter that makes us all individuals sharing the same form. What is the best definition of "man"? Aristotle and St. Thomas define man by saying we are rational animals. The rational part includes our will and our intellect, which we get from our soul. The animal part corresponds to the material life, including our senses and ability to move. All differences in skin tone, athletic ability, mathematical ability, etc., are contingent accidents. This demonstrates that all humans are children of God made in His image and likeness. Like a Divine Sculptor, He has formed us into images of Himself.

As we discuss the physics and chemistry of the material world, we will often overlook that the matter has a form and generalize to how matter behaves under our investigation. This does not mean that the form is not important, but we are simplifying things to better understand the matter's behavior.

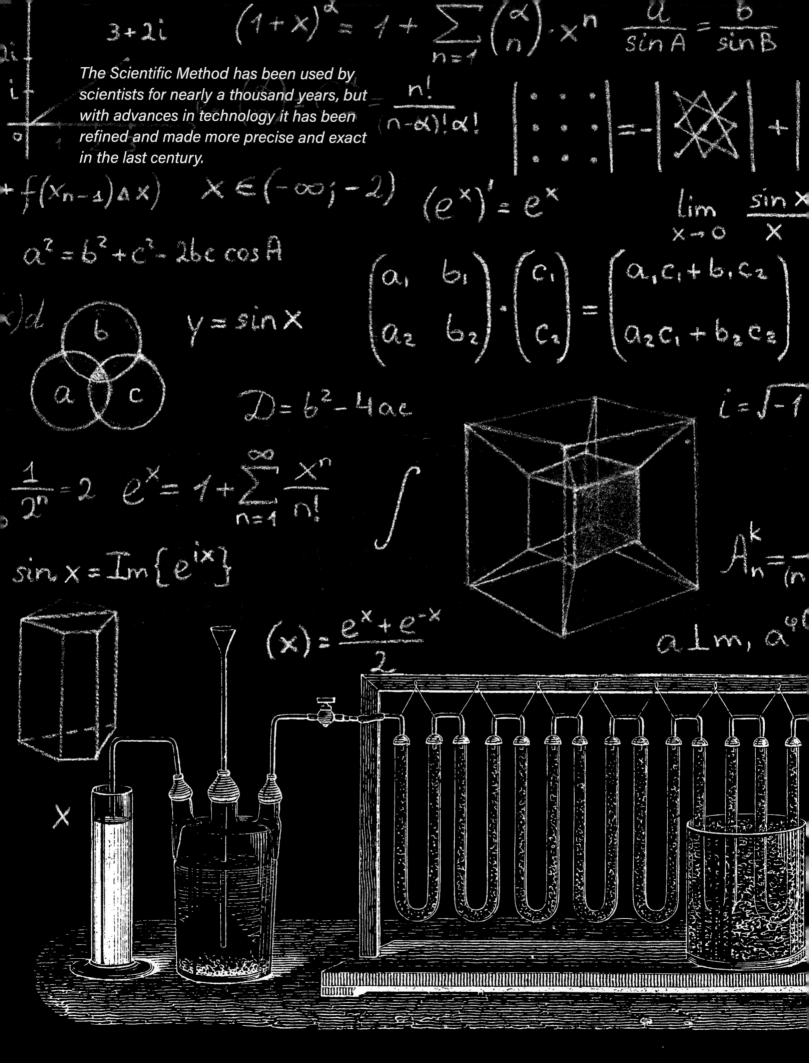

$3+2i$

$(1+x)^{\alpha} = 1 + \sum_{n=1} \binom{\alpha}{n} \cdot x^n$

$\dfrac{a}{\sin A} = \dfrac{b}{\sin B}$

The Scientific Method has been used by scientists for nearly a thousand years, but with advances in technology it has been refined and made more precise and exact in the last century.

$\dfrac{n!}{(n-\alpha)!\,\alpha!}$

$\begin{vmatrix} \cdot & \cdot & \cdot \\ \cdot & \cdot & \cdot \\ \cdot & \cdot & \cdot \end{vmatrix} = -\begin{vmatrix} \boxtimes \end{vmatrix} +$

$+ f(x_{n-1}) \Delta x)$

$x \in (-\infty; -2)$

$(e^x)' = e^x$

$\lim_{x \to 0} \dfrac{\sin x}{x}$

$a^2 = b^2 + c^2 - 2bc \cos A$

$\begin{pmatrix} a_1 & b_1 \\ a_2 & b_2 \end{pmatrix} \cdot \begin{pmatrix} c_1 \\ c_2 \end{pmatrix} = \begin{pmatrix} a_1 c_1 + b_1 c_2 \\ a_2 c_1 + b_2 c_2 \end{pmatrix}$

$y = \sin x$

$D = b^2 - 4ac$

$i = \sqrt{-1}$

$\dfrac{1}{2^n} = 2$

$e^x = 1 + \sum_{n=1}^{\infty} \dfrac{x^n}{n!}$

$\int$

$A_n^k =$

$\sin x = \operatorname{Im}\{e^{ix}\}$

$(x) = \dfrac{e^x + e^{-x}}{2}$

$a \perp m, \; a$

$x$

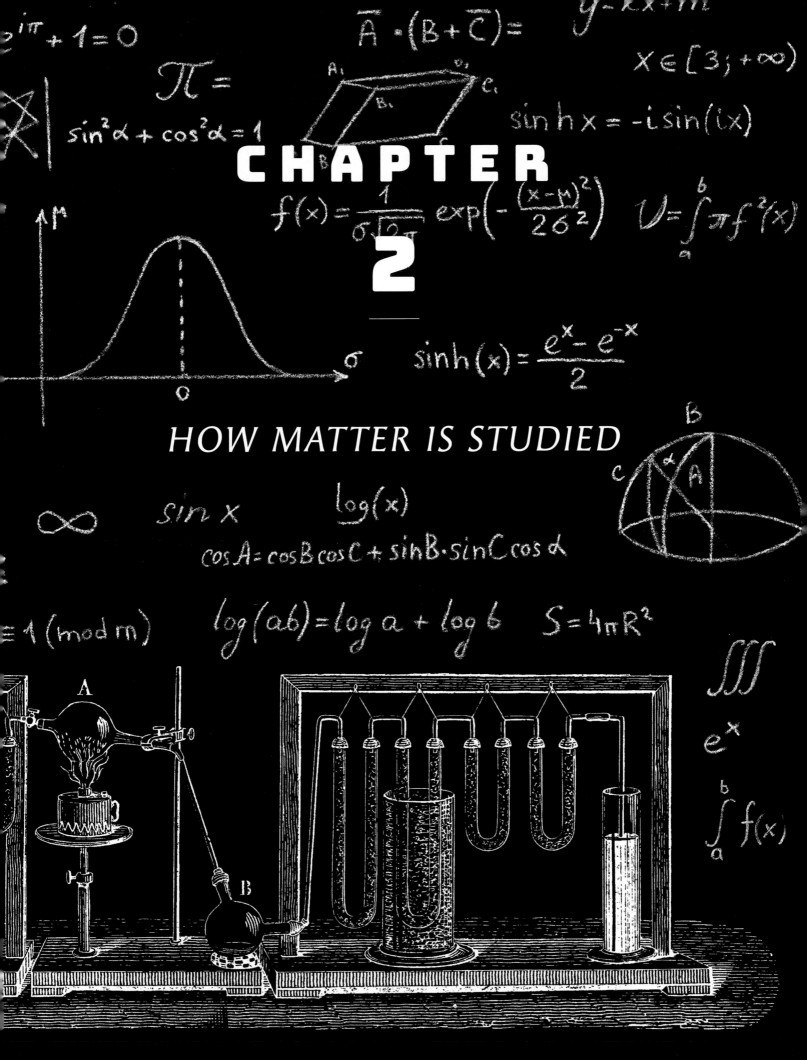

# CHAPTER 2

## HOW MATTER IS STUDIED

## THE SCIENTIFIC METHOD

The purpose of all sciences is to learn truth about a particular subject. The most general definition for **science** is that it is a methodological search for truth. The highest of the sciences is theology because it is the study of Him who is the highest subject of study: God. The second highest science is philosophy as it studies the created world. What we now call "science" was known as natural philosophy as it searches for truth about the *material* world, in other words, the stuff made of matter. Since the subject of study is material, the method of study is especially suited for the study of the material world.

The method science uses to study the material world, to study matter, is called the **scientific method**, and it consists of several steps. The number of steps and what they are called can vary depending on whom you are talking to, but generally they are:

1. Observation
2. Measurement/Research
3. Experiment
4. Hypothesis Formulation
5. Testing/Analyzing of the Hypothesis
6. Modification of the Hypothesis (if necessary)

Let's explore the winding journey scientists take through this method.

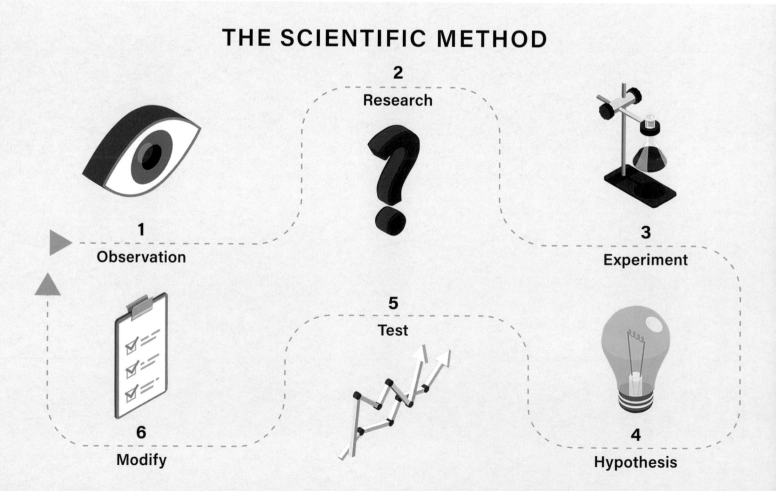

# THE SCIENTIFIC METHOD

**2**
**Research**

**1**
**Observation**

**3**
**Experiment**

**5**
**Test**

**6**
**Modify**

**4**
**Hypothesis**

## OBSERVATION AND MEASUREMENT

All human knowledge and all scientific pursuits begin with our senses. Our intellect makes use of the sense information we have to learn things about given objects. This is how we observe things, through our senses, and it is the first step to gaining knowledge. In the previous chapter, we mentioned some of the types of observations made, things like color, shape, size, how fast things move, or how things change in size. In all these observations, we are measuring things in one way or another, and these particular observations are **qualitative measurements**, meaning they give us a description using words.

Let's say a farmer has five watermelons to sell at the same price. We want to buy the biggest watermelon because then we know we have gotten the most for our money. How would we go about figuring out which one is the biggest? Well, we can lift the watermelons and compare their weight, but it might not be perfectly clear which one was the heaviest. We could set each one on a scale and, like a seesaw, measure how many equally weighted rocks are needed to balance the weight. The biggest watermelon would be the one that balances

*A quantitative method of measurement—like how many stones a watermelon weighs—would be more exact than a qualitative measurement, which only uses words to describe it.*

the highest number of rocks. The first way of comparing would be qualitative—we would just use words to tell someone, "I think this one is heavier"—while the second method using the number of stones we call **quantitative measurement**, as we have a *quantity* (a fixed number) used to compare a unit of measurement, in this case the number of rocks. The quantitative method would be more exact because we can rely on an outside unit of measurement (rocks) to give us more precise figures.

Humans have used measurements in trade for many, many centuries. In England, people used human scale measurements: number of feet between distances, number of strides to measure larger measurements (the yard), number of hands to measure how tall a horse is (the unit of hands), the amount of land an ox and one man could plow in one day (the acre). Of course, not everyone has feet of the same size or uses the same size stride when walking. And what if someone had a sick ox instead of a healthy one? This could lead to inaccurate data. There had to be some standard measure of what a particular length was.

Let us say the king's government makes a stick and calls it the "yard stick," and that a yard contains exactly three feet and each foot has exactly twelve inches. We now have a standard system of measurement used throughout the king's dominion. We call this a "system of measurement," where we have a number and a unit in which the measurement is made. We can say that an American football field is 120 yards long (including the end zones). In professional baseball, the pitching rubber on a baseball field is 60 feet 6 inches from home plate. A quantitative measurement gives a quantity (a number) in an agreed unit of measurement.

So far, we have only discussed units of length. We can also have units of volume—such as the fluid ounce, the gallon, the quart, a barrel—and units of mass (weight)—such as the pound, the gram, the kilogram, etc. Scientists across the world have all decided to use the same system of units of measurement since this makes communicating their findings with other scientists most understandable. This international system of physical units is called the **SI System** (SI = International System). It provides a defined and agreed upon unit for all the basic measurements scientists make: mass, length, time, volume, etc.

## Understanding Proportionality

Scientists and mathematicians say that a quantity is a function of something else when there is a possible mathematical relation that can be written. For example, the amount your parents pay for gasoline at the gas station is a "function" of the number of gallons they get; that is, they pay a set amount per gallon of gas they put in their car. The more gallons they get, the more they pay.

There are two types of functions common to science. The first type is when the quantities are *proportional* to each other. This means that both increase together or both decrease together, as we just demonstrated in that example of purchasing gasoline. The proportionality becomes a function with the use of a constant. In the example of the cost of gasoline, the proportionality constant is the price of the gasoline. If gasoline is $3 per gallon, and we buy 5 gallons, we pay $15. If we buy 10 gallons, we pay $30, and so on.

The other relation is called *inverse proportionality*. In this case, as one quantity goes up, the second one decreases. They move *opposite* to one another. An example of this relation would be the size of cake slices split among a group of people. If you want to split a cake equally, then the larger the group, the smaller each piece has to be. The smaller the group, the larger each piece can be. So we can say the piece size and the number of people are inversely related. This is an important concept to grasp in math and science.

INTERNATIONAL SYSTEM OF MEASUREMENT

Kilogram
kg

Candela
cd

Kelvin
K

Meter
m

Second
s

Ampere
A

Mole
mol

## The Canon of Scripture

In a sense, a measurement is a judgment on an observation using a standard unit. The Greek work *kanon* means a measuring stick. The word was used to denote a standard or rule. In the discussion about which books were to be included in the canon of scripture, *canon* took the meaning of a defined collection. Not all the books that were held as good Christian literature were included in the canon. What were the standards used to determine the canon? The book had to be orthodox, universal, and liturgical. This means that the books could not teach anything opposed to the true Faith once given to the saints. It had to be found everywhere in the Church, and it had to be used in the Church's liturgy. There were very orthodox books that did not find their way into the canon because they failed to meet one of these criteria.

*Papyrus 46* ▶
*Containing 2 Corinthians 11:33-12:9*

# FORMULATING A HYPOTHESIS AND EXPERIMENTING TO TEST IT

Scientists strive to make qualitative observations along with careful quantitative measurements as they seek knowledge. These observations and measurements are the beginning of the method used in studying the phenomena of the natural world. But once these are made, one might wonder how it is possible to understand what is happening. This is something we do naturally as it is proper to the human intellect (other animals make no such studies). The scientist strives to make this understanding based solely on material causes.

For example, say you are looking for a container to keep coffee warm—you would obviously want to find a container that kept it the warmest for the longest amount of time. You have noticed that covered Styrofoam cups keep coffee warm longer than uncovered cups or cups made of cardboard, glass, or paper. But this is just something you observed—you don't know for sure. So, you place hot water in different cups and then measure (quantitatively) the temperature of the water using a thermometer every five minutes. Perhaps you take it a step further and notice that, of the two Styrofoam cups you used, one is thicker, and it quantitatively keeps water warmer the longest. You can now make the educated guess that as the thickness of the Styrofoam increases, the coffee stays warm longer. This educated guess is a **hypothesis**: a proposed explanation for an observed phenomenon. A good hypothesis must be made in such a way that it is possible to test it using material means, and it must be possible to show that it is incorrect. The hypothesis should also extend the observation in such a way as to predict something that is yet to be observed. This prediction can be used to test the hypothesis just formulated.

Once a hypothesis is formulated, the scientist begins a more formal effort to conduct an experiment and test the hypothesis, which includes inviting others in his or her field to experiment and test. The experiment is set up in such a way that we can quantitatively measure something that can be used to confirm or disprove the hypothesis. Throughout the experiment, it is imperative to simplify things in order to concentrate on the question at hand. In the example of the Styrofoam cups, the scientist would control the amount of water (always using the exact same volume of water) as well as the initial starting temperature of it. The cups should be the same shape as well, except for their thickness. The point is to unify all the variables (make them the same) except the variable you are testing (the thickness of the Styrofoam).

If the experiment fails to show that thicker Styrofoam does indeed keep water warmer for longer, the scientist may modify the hypothesis, meaning change it, and then conduct more experiments to test the new hypothesis. But, if the experiment shows the hypothesis to be true, the scientist can start asking other questions: Does increasing the thickness of a glass cup have the same effect of keeping the liquid warm longer? Does changing the shape of the cup change the rate of cooling of the water? What if I have a liquid other than water in the cup? Each of these questions provides a new hypothesis to be tested.

Once a collection of experiments in a field have failed to show a hypothesis to be incorrect (that is to say, the experiments repeatedly support the hypothesis), it can be said that we have a theory to explain the phenomena. A **theory** is a systemic explanation of a natural phenomenon that can predict the results to be observed in a situation. In our example, our theory of heat transfer states that the thicker the Styrofoam, the slower the cooling of the coffee. This was obviously the same as our hypothesis, but it is important to note the difference between a hypothesis and a theory. The former is an educated guess based on observations and initial measurements; the latter is supported by repeated experimentation and evidence. That being said, our theory is not a proven fact. It can always be shown to be incomplete or incorrect by later scientists doing better experiments. Let's close out this chapter by moving past theories and into the area of laws.

## LAWS OF NATURE

A **law of nature** differs from a theory in that it cannot change with time—it has been proven to be constant. The laws scientists speak of, such as the law of gravity, give a mathematical explanation to observed phenomena; in other words, they give a mathematical form to the results of careful observation and

**Remember:**

*If two quantities are inversely proportional, that means as one goes up, the other goes down. This is the opposite of two quantities being proportional (when one goes up or down, the other also goes up or down).*

experimentation. This mathematical relation is always called a **function**. We can say the money one spends in buying apples is a function of the price for each pound of apples. If the price for apples is $2 per pound, then we would pay $10 for five pounds of apples (2 x 5).

In returning to the law of gravity, we all know that if something is dropped it will fall to the ground. Sir Isaac Newton gave a math equation for this relation, which we call the law of universal gravitation. His equation says that the force of gravity changes with the masses of two objects and the distance between them. In mathematical terms, we would say that gravity is a function of the two masses and the distance between them. As we march through this book, we will explore other laws like this one.

With our understanding of the scientific method behind us, let us move on to the concept of motion and the forces behind it.

## FOUNDATIONS REVIEW

✓ Since the subject of the empirical sciences is material, the method of study is especially suited for the study of the material world. The method science uses to study the material world is called the scientific method, and it consists of several general steps: (1) observation, (2) measurement/research, (3) experiment, (4) hypothesis formulation, (5) testing/analyzing of the hypothesis, (6) modification of the hypothesis (if necessary).

✓ In the scientific method, we observe the phenomena of the world using our senses and then make qualitative observations (using words to describe things). We can then take quantitative measurements based on agreed upon units of measure, and from this observation and measurement form a hypothesis: an educated guess and proposed explanation for an observed phenomenon. Once we have our hypothesis, we test it through repeated experimentation to either prove or disprove it.

✓ When we have completed our experiments and determined the viability of our hypothesis, we can create a theory, which is a systematic explanation of a natural phenomenon that can predict the results to be observed in a situation. A theory, however, can be shown to be false by later experiments. A more constant and mathematically proven declaration in science is called a law of nature, such as the law of gravity.

# The Four Causes

In this chapter we discussed the modern scientific method which is used to learn about the workings of the natural world. This method is limited to observation and experimentation; as such, the modern scientific method is only useful to know how to make or change things. It is not meant to give us positive knowledge of a thing's form and purpose. A full knowledge of a creature includes these elements.

The philosophical framework that comes to us from Aristotle and St. Thomas Aquinas states that to fully know something there are *four causes* that must be known—meaning four things that make it what it is. These are:

1. the material cause,
2. the efficient cause,
3. the formal cause, and
4. the final cause.

The *material cause* tells us what the creature is made of—its material components. It tells us the quantity and type of matter it contains. The *efficient cause* tells us how that thing comes into existence and how it remains in existence. The scientific method chooses to limit its focus to these two causes—it does not address formal and final causes. In fact, many of the early scientists (Descartes, Newton, etc.)

stated that there were no such things as formal and final causes. Philosophers have said otherwise.

The *formal cause* states what makes the object be what it is, while the *final cause* is the purpose of the thing; it tells us what the thing is for. Things made by human hands always have a purpose. If asked what a table is, you are likely to include the purpose of a table as part of your answer; likewise, if you were asked to define a knife.

Within living organisms, it can be relatively easy to know the final cause of something. For example, an eye is for seeing. For nonliving things, this final cause may not be as easy. But this does not mean we can ignore the final cause of something or say that final causes do not exist.

What is our final cause, we rational animals? The Baltimore Catechism states God made us to: "Show His goodness and make us happy with Him in heaven." Our final cause is to share the life of God with Him for all eternity in heaven. This is true happiness. We are creatures who are meant to be happy with God.

An object's speed is a measurement of how fast it moves across a given unit of distance. Bullet trains are some of the fastest machines in the world, with speeds of almost 300 mph!

# CHAPTER

## 3

# MOTION

*Speed, Velocity, and Momentum*

## MOVING FROM HERE TO THERE

The modern study of the natural world begins with the field of **mechanics**, which is the study of the movement of objects in space. This was of interest to early physicists as they wanted to be able to have a simple model to understand the movement of the planets as observed from Earth. The Aristotelian understanding of the movement of objects is that they only moved while something was acting upon them. For example, a toy car moves only while it is being pushed or pulled by your hand. When the act of pushing or pulling ends, then the toy car will slow down and then stop moving. This observed understanding of motion was the accepted view until a transition occurred between the fifteenth and seventeenth centuries.

The next few chapters in this book will examine the new way of looking at how objects move. We will begin with the study of how we quantify the movement of objects through space, which is called **motion**. Once we understand how we measure motion, then we will be ready to study changes in an object's movement.

## SPEED

The study of motion is called **kinematics** (one of the neater sounding words in science!). As you might expect, the simplest way to look at an object's motion is by measuring how fast it is moving. Just think of a race you've had with friends where you run from one point to another to see who can get there first.

*Engineers study for years and spend millions of dollars just to shave a few seconds off a racecar's time around the track.*

To gauge how fast an object is moving we need two measurements: (1) the distance it moves and (2) the time it takes for it to cover that distance. We can measure the distance the object moves in feet, meters, miles, or kilometers, and the time it takes the object to cover the distance in seconds, minutes, or hours.

This measure is called **speed** by physicists. Speed is calculated by dividing the length of the object's motion—inches, yards, meters, miles, etc.—by the time it took for it to cover that distance—seconds, minutes, hours. The units of speed you are probably most familiar with are either "miles per hour"

*The cheetah is the world's fastest land animal, reaching speeds of up to 70 mph!*

(abbreviated mph and used in the U.S.) or "kilometers per hour" (abbreviated kph and used most everywhere else in the world), as these are used to measure the speed of cars, trains, and airplanes. The most common unit scientists use to measure speed is meters per second: $^m/_s$ (giving the units as a fraction like this is common in the natural sciences).

The calculation of an object's average speed should not be difficult. Let's go through an example.

Say you complete a 10-mile hike in 5 hours. We would calculate the average speed of the hike by dividing the distance (10 miles) by the time (5 hours) to get a 2 miles-per-hour speed, which is also written as 2 $^{mi}/_{hr}$. If a sprinter runs the 100-meter dash in 10 seconds, we calculate his average speed by dividing the 100 meters by the 10 seconds to get 10 $^m/_s$. Notice that in both examples we looked at the *average* speed. Why would we do that?

If you think about a hike, you sometimes walk at an even pace. At other times, you may see something interesting up ahead (a giant mushroom!) and so you run to look at it. Maybe at times you slow down to enjoy the scenery or stop to rest or have a drink. In all these cases, you are moving at different speeds at different moments of time. But even taking all that into account, we still get the average speed of 2 $^{mi}/_{hr}$.

As you can see, the average speed does not give us the full picture. The speed of an object at any particular moment in time is called the **instantaneous speed** (the speed at that "instant"). If we examine the 100-meter dash, the racers start at a full stop, meaning they have an instantaneous speed of 0 $^m/_s$. As the race starts, the runners increase their speed until they reach their maximum running speed. Most world-class sprinters cross the finish line near the highest instantaneous speed, at about 12 $^m/_s$. To think of that in a more familiar way, that means they are running 12 meters in a second, which is almost 40 feet!

Here we see an example of a change in an object's motion. The sprinters can increase their speed and decrease their speed. Colloquially, we would say that an object increasing its speed *accelerates*, as a car does when the light turns green, or *decelerates*, as a car should when it approaches a stop sign.

Speed is important when studying an object's motion, but it is not the only kind of measurement we can look at.

## VELOCITY AND ACCELERATION

Another aspect of an object's motion that is important to scientists in addition to speed is the direction the object is moving. The term **velocity** is used to describe the speed of an object while also factoring in the direction of its motion. So you might say that an airplane is flying at a constant speed of 400 mph in an easterly direction. This would be the velocity of the airplane. Or perhaps a car drives at an average velocity of 25 mph *away* from home or 25 mph *toward* home.

Generally speaking, scientists think of the movement of an object through space as motion through a grid. The center of the grid can be placed at any location of the scientist's choice, although they often center it in a position of interest. For example, when your family drives somewhere and leaves home, your house is the center, or starting point—this is the "position of interest." When an object moves to the right in the grid, physicists say the object is moving with a *positive velocity*. When the object moves to the left, they say the object has a *negative velocity*. This is one reason you will study negative numbers in algebra, as these are foundational to the mathematical description of the world. It is important to know that this is just a way of thinking about motion in a mathematical grid—it does not mean if you move to the left yourself you are moving "negatively" (backwards).

Unlike velocity, there are some things that have an **absolute quantity**, meaning they can never be negative. An example of this would be speed, as the lowest motion an object can have is to be still—a speed of 0 $^m/s$.

A moment ago, we noted that an object can change speeds—that it can accelerate (pick up speed) and decelerate (slow down). These are common terms. However, in physics **acceleration** is the technical term for the change of velocity with time. This means that an object accelerates when its speed changes, when the direction of movement changes, or when they both change. When you feel movement—for example, when you get on a rollercoaster—what you feel is the change in motion, or the acceleration of the object. When a car slows down or increases speed you can actually *feel* the change. But you may also feel the acceleration of an object whose speed does not change, like when a car makes a turn without changing speed. A better example is on an airplane that cruises at constant speed and then makes a turn. You didn't really feel the movement of the plane until it started to turn, despite the speed of the movement being no different.

*Physics Fun Fact:*

*On the previous page we learned cheetahs are quite fast. Perhaps surprisingly, though, if a cheetah were to race an ostrich for a longer stretch, the ostrich would win. While cheetahs can have quick instantaneous speeds but then tire out, the ostrich can run at over 40 mph for as long as thirty minutes! This means the ostrich maintains its speed for longer.*

## Cartesian Coordinate System

An important idea in mathematical understanding in physics was the creation of a grid in which physical information could be expressed, meaning something could be shown by graphing it. For example, if we took your weight and height, we could chart those two figures and place a dot that represents you on the chart (your doctor does this for you to measure your growth compared to other children). The expression of data by grids was developed by the French polyglot René Descartes. The grid is called the Cartesian coordinate system. The coordinate system starts at a chosen point called the "origin." Through the origin, we draw horizontal and vertical lines that cross. These lines are called **axes**. The horizontal one is called the x-axis, and the vertical one the y-axis. The lines are then divided into even steps starting at the origin. As we go right on the x-axis or up on the y-axis, the steps are given *positive* values. As we go left or down, then scientists use *negative* numbers to give the position. We can then use values for the x-position and y-position—the coordinates—for any point in the axis. These are written as: (x-value, y-value). This system of graphing data is an important concept to understand if we want to be scientists.

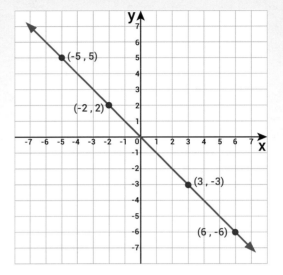

## TOTAL MOTION

So far we have talked about the motion of objects without concerning ourselves with any detail of what the object looks like. For the most part, scientists simplify the object of study as much as possible. The color, size, and shape of the moving object seem of little importance when investigating moving objects. In reality, some of those characteristics do play a part in the equation and we will discuss how that is in the next few chapters.

For our purposes now, we will focus on one characteristic that is extremely important when considering an object's motion, which is the mass of the object (remember from chapter 1: mass is the amount of matter in an object). Sir Isaac Newton stated that the total motion of an object includes its mass and velocity. Modern scientists call this **momentum**, which is calculated by multiplying the mass of the object by its velocity. Let's look at an example.

Say you have two shopping carts like you see at grocery stores: a regular adult-sized one, and a child's toy cart. Let's say the adult cart is 10 kg (about 20 pounds) and is moving at 2 m/s (about six feet per second), and the child's cart is 1 kg (about 2 pounds) but is also moving at 2 m/s. To help you see this better, we will write it like this:

| Adult Cart | Child's Cart |
| --- | --- |
| Weight: 10 kg<br>Speed: 2 m/s | Weight: 1 kg<br>Speed 2 m/s |

Taking into account these measurements, we would say the adult cart has more momentum, even though both carts are traveling at the same speed. The momentum of the adult cart is: 10 kg x 2 $^m/s$ = 20 $\frac{kg \times m}{s}$. Meanwhile, the toy cart's momentum is: 1 kg x 2 $^m/s$ = 2 $\frac{kg \times m}{s}$. You instinctually know this even without looking at the math, as a bigger object moving at the same speed as a smaller one is harder to stop (it has more momentum). It is also harder to *start* moving heavier objects from a stopped position, which is why a loaded truck takes a lot longer to get going when a stoplight turns green than a small car. Much of what we will study in this unit involves things you probably know from common sense but have never considered on a deeper level. That's what we are trying to do in making you junior scientists—make you think more deeply about the world and ask questions about why things are the way they are. We will continue our efforts in the next chapter by looking at the changes in motion of an object with a given mass.

## FOUNDATIONS REVIEW

✓ The simplest way to look at an object's motion is by measuring how fast it is moving. To gauge how fast an object is moving we need two measurements: (1) the distance it moves and (2) the time it takes for it to cover that distance. Speed is calculated by dividing the length of the object's motion—inches, meters, yards, miles, etc.— by the time it took for it to cover that distance—seconds, minutes, hours (for example, 35 mph). Speed can change over time as an object accelerates (picks up speed) or decelerates (slows down).

✓ In addition to speed, another aspect of an object's motion that is important to scientists is the direction the object is moving. The term velocity is used to describe the speed of an object while also factoring in the direction of its motion. So you might say that an airplane is flying at a constant speed of 400 mph in an easterly direction.

✓ One characteristic that is extremely important when considering an object's motion is its mass (the amount of matter in the object). Sir Isaac Newton stated that the total motion of an object includes its mass and velocity. Modern scientists call this momentum, which is calculated by multiplying the mass of the object by its velocity.

# Transubstantiation

Any material thing has the potential to be changed. This change must be brought about by an outside mover. A block of marble has the potential to be turned into a statue, but this must be actualized by something outside itself. A chisel cuts into the marble to shape it. The chisel is moved by the hand of a man. The man has the capacity to sculpt as part of his human nature. As you see, we just keep adding layers to the understanding of the motion of the marble to the statue. *At some point, we must come to a point where a mover causes motion without His being moved.* Aristotle called this the "unmoved mover." St. Thomas Aquinas said, "That is which we call God." God is the final efficient cause of all motion, as He Himself is unchanging.

For most of our experience, the form of a thing agrees with its essential accidents (see insert following chapter 1). Wine is grape juice that has been fermented to turn it into wine. This requires the use of yeasts which take the sugars and convert it to alcohol. The flavor of the wine is different from that of the juice, the density is different, the smell is different, and even the color may change. We use our senses to determine whether we have juice or wine.

But there is one exception to this. At the moment of the consecration of the Holy Sacrifice of the Mass, the host and the wine are converted into the Body and Blood of our Lord Jesus Christ. The accidents of bread and wine remain—meaning they still seem like bread and wine to our senses. The host tastes, smells, and feels the same before and after the words of consecration. The consecrated wine also has the same accidents as the wine before the consecration. So what is the change?

Remember that God is the author of all creation. During the Mass the priest acts in the person of Christ; he is the priest and sacrifice. As such, when the words "This is my Body" are spoken, it is the Second Person of the Holy Trinity who speaks and changes the form/substance of the bread (what it actually is) into His Body, Blood, Soul, and Divinity. Similarly, the wine in the chalice is changed into our Lord's Blood. This most Holy Sacrifice is offered at all Catholic Masses for the salvation of the world. The change from bread and wine to the Body and Blood of Jesus is called transubstantiation (trans = change, so "change of substance").

The Newton's Cradle is a peculiar device that consists of a row of swinging spheres or balls, whereby the balls on either end swing back and forth and knock the other, transitioning a force through the stationary spheres between them. It is used to demonstrate a basic tenet of physics, that of the conservation of momentum and energy.

# CHAPTER
## 4

## *NEWTON'S LAWS OF MOTION*

## WHAT CAUSES MOTION?

In the previous chapter, we looked at how motion is quantified, or measured. We will now discuss the way changes of motion are understood and modeled. It is not always enough to simply look and notice something—sometimes we must seek to understand why something is happening, and then formulate a way to express what we have learned.

One of the first physicists was Aristotle. He gave an explanation of the motion of an object based on his observation as it matched his experience. Think of that toy car we mentioned in the last chapter. It rolls on the table while you are pushing it, but when you remove your hand from the car, it slows down until eventually it will stop moving altogether.

From such observations, Aristotle stated that motion required an "agent" to cause the motion, and when that went away the motion stopped. In the case of the toy car, you are the agent moving it with your hand.

But the planets and other objects in the sky seem to have self-movement, as there is nothing acting upon them. This implied that the planets were either moved directly by a divine mover or are themselves divine in some sense. This explanation did not fit within the type of explanation that could be given using the scientific method, so a completely new framework was proposed.

## INERTIA

A change in the way we understood motion started in the fifteenth century in Catholic universities, but was given its final form by Galileo Galilei. Galileo was a famous Italian astronomer, physicist, and engineer of the sixteenth and seventeenth centuries. He helped develop a new framework for how we understand motion. This new framework looked at an object's motion as if it were constant and unchanging until something changed its motion. What do we mean by that?

Well, if an object is not moving, it would remain like that until something comes along and changes that, right? This should make sense. A soccer ball sitting on the field will remain there until something happens to make it move, presumably a player coming to kick it or perhaps a very strong wind blowing it. This fits with our experience.

The second part of our understanding is that an object in motion will retain that motion unless something changes it. This would imply that once the soccer ball is kicked, as long as nothing gets in its way, it should continue to roll forever. The fact that it does not means that we need to look at other sources that can act upon it to slow it down, change its direction, or stop the ball altogether (as we know it will eventually stop).

GALILEO GALILEI

Scientists say that objects have **inertia** when they are in motion. This means that they will keep the motion they have until something comes along and changes it. An object's inertial motion is related to its quantity of matter and its mass. An object with *more* mass will have *more* inertia than one having less mass. Think back to the discussion of our two shopping carts from the previous chapter, the bigger one and the toy cart. The bigger one, which had more momentum (and more mass) would be harder to bring to a stop. Likewise a coal train has more inertia than a compact car, as it has more mass. In the simplest terms, we could say that the larger and heavier an object is, the harder it will be to stop. Its inertia is greater. The concept of inertia tells us that changing the motion of an object is more difficult the greater its mass.

*Since the adult grocery cart (with all that's in it) has more mass and matter than the child's cart, this means its inertia will be stronger, meaning it will be harder to stop its forward motion once it gets moving.*

## Sir Isaac Newton

Isaac Newton was one of the world's greatest and most influential physicists and mathematicians. He was born on Christmas Day (on the Julian calendar) in 1642 in Lincolnshire, England. He received a classical education with a strong foundation in mathematics. He was accepted into Trinity College, Cambridge, in 1661. Newton developed the branch of mathematics called calculus (which Gottfried Wilhelm Leibniz also developed independently in Germany) to be able to work out his theory of mechanics. His driving force was to give a simple mathematical model of the elliptical orbits of the planets around the Sun. His most important work in mechanics is the *Philosophiae Naturalis Principia Mathematica* published in 1687. In this work he set out his laws of motion and the law of gravitation and used the principles of calculus to give the elliptical orbits of the planets. He also researched optics, the study of the behavior of light as it reflects and moves through media.

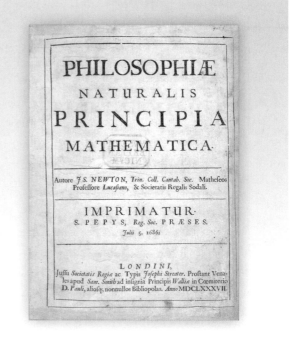

# NEWTON'S LAWS OF MOTION

Our complete modern understanding of motion was proposed by Sir Isaac Newton. We have mentioned him a few times already, and it is from him that this chapter gets its name. Isaac Newton was an English mathematician, physicist, and astronomer living in the seventeenth and eighteenth centuries. He is remembered today as one of the most influential scientists in history, primarily because he proposed three laws that form a predictive model for the motion of objects—this means the laws help us predict how objects move. Newton's proposal completely changed the way scientists look at the physical world.

When we begin to study these laws, it is important to note that the idea of motion Newton begins with—an assumption he makes—is that an object's normal motion is in a straight line at a constant speed. In other words, with all things being equal and no outside forces acting upon an object, there is no reason an object should turn and change directions or speed up *on its own.* We can say that it has unchanging velocity. Any motion that deviates (changes) from a straight line is not inertial motion, meaning that the object is changing direction and something is causing this change of direction. Remember that changing an object's direction even without changing the speed is considered acceleration of the object.

Considering all this, Newton's **first law of motion** states that an object in motion will continue in motion and an object at rest will continue at rest unless an outside force acts upon it. If an object's motion is changed, then something outside of it caused the change. The action of changing an object's motion is called a force, which is a word we have already used a few times and you are probably already familiar with it. A **force** is defined as any pull or push that changes an object's motion. With the toy car, your hand was the force moving it. Forces act on the object only as long as the force is directly applied to the object (when you remove your hand, the force is no longer being applied).

A game of baseball is a good way to understand forces and changes in velocity. The baseball is accelerated by a pitcher only as long as the ball is in contact with his hand. But if the batter hits it, the bat changes the direction of the baseball, though only for the moment that the ball is in contact with the bat. After the force of the pitcher or bat stops acting on the ball, then other forces act on the ball as it changes from a straight-line motion (perhaps the wind affects it or an outfielder catches it and stops the acceleration of the ball). A force that acts on the object for a short time like this is called an **impulse.**

*The phenomenon of a bat striking a baseball gives us a perfect example of physics at work, specifically with forces and changes in velocity of objects.*

## FORCES ON A BASEBALL

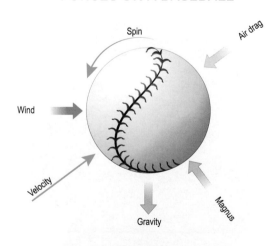

Newton's second law of motion is related to the effect of a force on the object's motion. The **second law of motion** states that the acceleration of an object of constant mass (like a baseball that doesn't change size) is proportional to the force applied on the object. This means that the larger the force applied to a given object (the harder the pitcher throws it or the batter hits it), the greater the acceleration on the object. You could also say that if you desire a greater acceleration of the object, a greater force is needed to cause the acceleration. This is why bat speed (how quickly the batter swings the bat) is a key factor in the success of baseball players, determining those who hit well and have many homeruns. The faster you swing the bat, the farther the ball will go.

This phenomenon is seen mathematically in the following equations, where *F* stands for a force, *m* is the mass of the object, and *a* is the object's acceleration.

$$F = m \times a \ \ or \ \ a = F/m$$

In the standard units used by scientists, the mass is in kilograms, the acceleration in meters per second per second, and the unit for force is called the Newton (N), named after Isaac Newton (maybe one day you will have a unit of measure named after you!). The first equation would be used to calculate the force when we know the acceleration and the mass. The second equation would be used to calculate the acceleration from a given force and mass.

# NEWTON'S SECOND LAW OF MOTION
### (PUSHING A CAR AND A TRUCK)

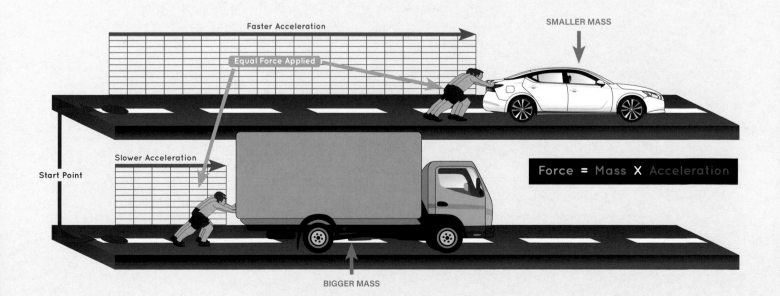

**Reaction**

**Action**

We could plug in numbers to give examples, but it can start to look complicated, especially for children your age. The important thing to understand is just the thinking behind it, rather than memorizing the formulas. So simply remember that the more force an object has applied to it, the greater its acceleration is.

Finally, the **third law of motion** is known as the action-reaction law. It states that when a force is applied to an object, the object applies a *reaction force* equal in size to the force applied on it—but in the opposite direction. This means that as a chandelier pulls down on the stud in the ceiling, the ceiling is applying an equal force as it pulls up on the chandelier. Or, as you push on the wall, the wall "pushes back" at you with the same force.

A common example of the action-reaction law is what happens when someone tries to step from a boat onto a pier. As he pushes on the boat to walk forward, the force of his step exerts an equal force in the opposite direction, causing the boat to move away from the pier. This can create a difficult situation and a feeling of panic as he nearly falls into the water! Another example is the "kick" of a rifle on a hunter's shoulder. As the force pushes the bullet out of the barrel, a reactive force pushes the rifle into the shoulder of the hunter. If the hunter does not push with an equal force, then he will be pushed backward by the rifle.

In our next chapter we will continue our discussion of forces, getting into the specific kinds of forces that we see in the world.

## FOUNDATIONS REVIEW

✓ For something to be moving, it must have an agent acting upon it. This means there must be something causing it to move. If the cause of the motion is removed, the object will no longer move.

✓ Scientists say that objects have inertia when they are in motion. This means that they will keep the motion they have until something comes along and changes it. An object's inertial motion is related to its quantity of matter and its mass. An object with *more* mass will have *more* inertia than one having less mass.

✓ Sir Isaac Newton formulated three laws of motion. Newton's first law of motion states that an object in motion will continue in motion and an object at rest will continue at rest unless an outside force acts upon it. The second law states that the acceleration of an object of constant mass is proportional to the force applied on the object. And the third law is known as the action-reaction law and states that when a force is applied to an object, the object applies a reaction force equal in size to the force applied to it—but in the opposite direction.

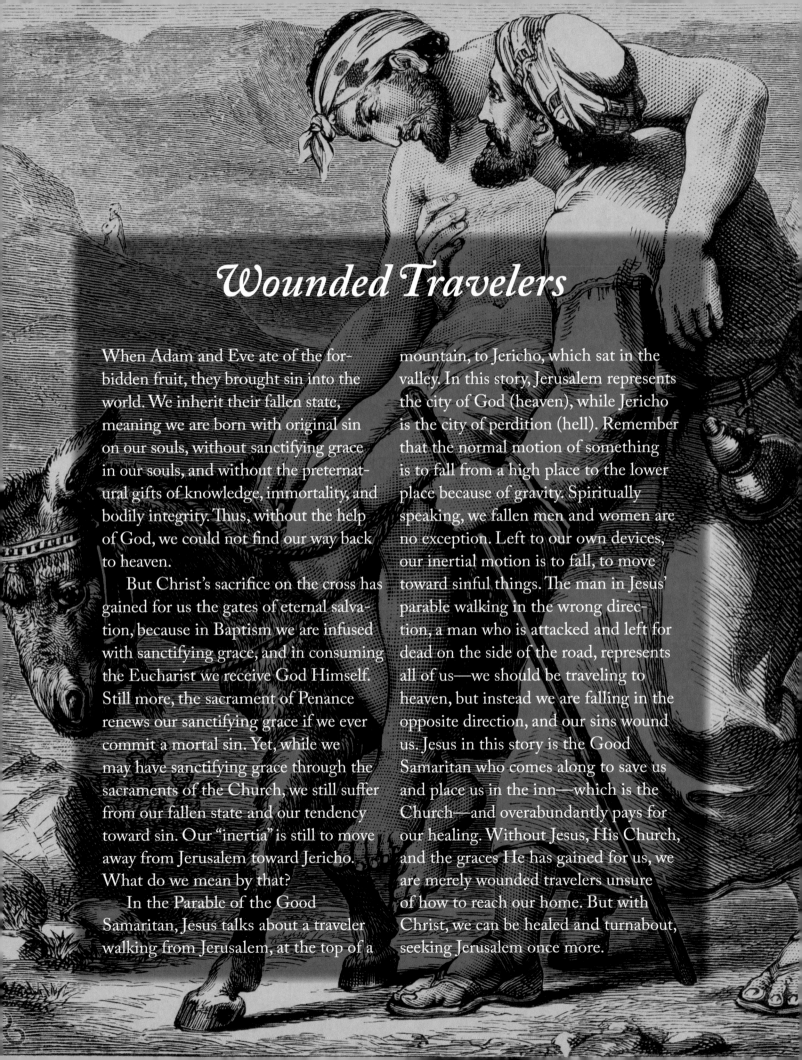

# Wounded Travelers

When Adam and Eve ate of the forbidden fruit, they brought sin into the world. We inherit their fallen state, meaning we are born with original sin on our souls, without sanctifying grace in our souls, and without the preternatural gifts of knowledge, immortality, and bodily integrity. Thus, without the help of God, we could not find our way back to heaven.

But Christ's sacrifice on the cross has gained for us the gates of eternal salvation, because in Baptism we are infused with sanctifying grace, and in consuming the Eucharist we receive God Himself. Still more, the sacrament of Penance renews our sanctifying grace if we ever commit a mortal sin. Yet, while we may have sanctifying grace through the sacraments of the Church, we still suffer from our fallen state and our tendency toward sin. Our "inertia" is still to move away from Jerusalem toward Jericho. What do we mean by that?

In the Parable of the Good Samaritan, Jesus talks about a traveler walking from Jerusalem, at the top of a mountain, to Jericho, which sat in the valley. In this story, Jerusalem represents the city of God (heaven), while Jericho is the city of perdition (hell). Remember that the normal motion of something is to fall from a high place to the lower place because of gravity. Spiritually speaking, we fallen men and women are no exception. Left to our own devices, our inertial motion is to fall, to move toward sinful things. The man in Jesus' parable walking in the wrong direction, a man who is attacked and left for dead on the side of the road, represents all of us—we should be traveling to heaven, but instead we are falling in the opposite direction, and our sins wound us. Jesus in this story is the Good Samaritan who comes along to save us and place us in the inn—which is the Church—and overabundantly pays for our healing. Without Jesus, His Church, and the graces He has gained for us, we are merely wounded travelers unsure of how to reach our home. But with Christ, we can be healed and turnabout, seeking Jerusalem once more.

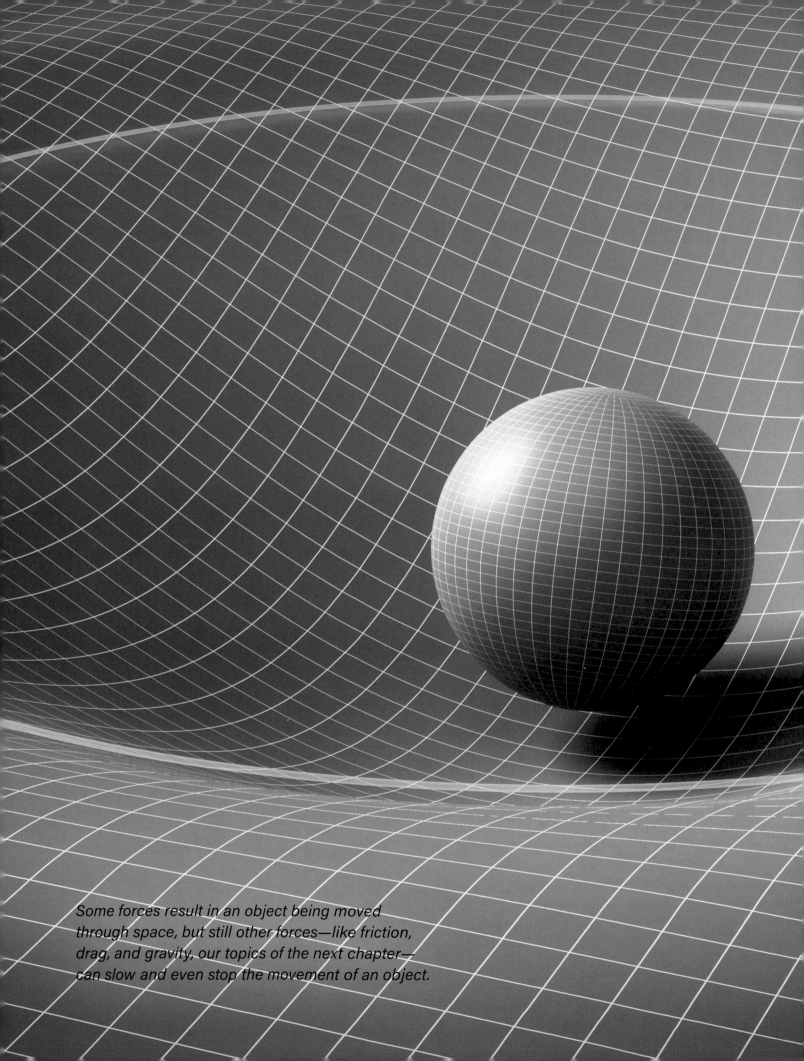

Some forces result in an object being moved through space, but still other forces—like friction, drag, and gravity, our topics of the next chapter— can slow and even stop the movement of an object.

# CHAPTER
## 5

### FORCES
*Friction, Drag, and Gravity*

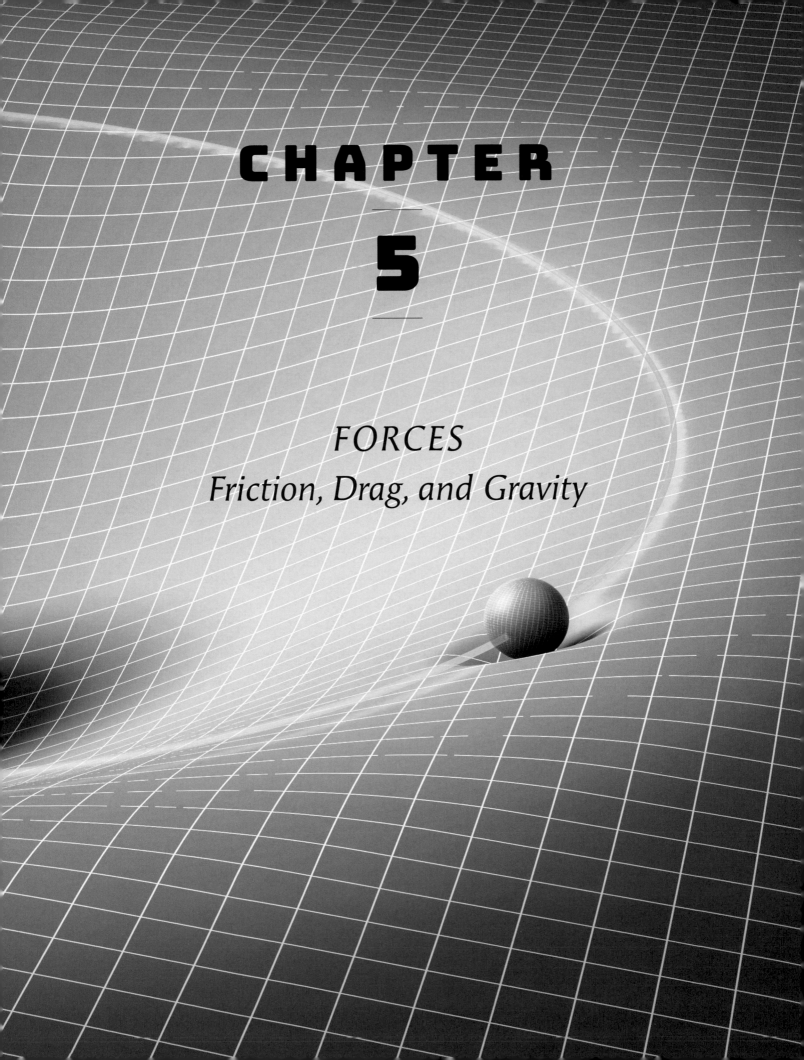

## KINDS OF FORCES

In the previous chapter, we discussed the Newtonian understanding of motion in contrast to the previous Aristotelian framework. The Aristotelian framework seems to better describe what we observe when we see objects that move only when moved by a mover (you pushing that toy car) but that stop moving when the mover stops moving them. We may struggle to make sense of the Newtonian idea of straight inertial motion. How could something move when it appears as though nothing is moving it (like the planets around the sun we mentioned at the start of the last chapter)? It is a difficult thing to make sense of.

Nonetheless, when we combine Newton's three laws of motion, we get a very powerful predictive system to understand and manipulate the motion of objects. We will do that in this chapter by looking at three forces that change motion as the object moves or is at rest: friction, drag, and gravity.

As we start looking at many forces acting on an object at the same time, we have to keep in mind that when we have equal and opposite forces, there is no "net change" to the object's velocity. The object will accelerate only when a given force overpowers another force. Think of the seesaw—if you and a friend who weigh exactly the same are sitting on either side, it will balance perfectly. There will only be a change and one side will overpower the other if one of you picks up a heavy rock to tip the scales.

When we started discussing motion, we talked about the speed, velocity, and acceleration of a given object without taking any of the properties of the object into account. But then we added the mass of the object to give a better description of its total motion. An object's mass is the main measure of its inertia and is used to calculate the effect of a force on the acceleration of the object. As we look at other forces, the shape and material of the object become important. With that in mind, let's discuss our first kind of force.

*Remember:*

*An object's mass is the main measure of its inertia. It is used to calculate the effect of a force on the acceleration of the object.*

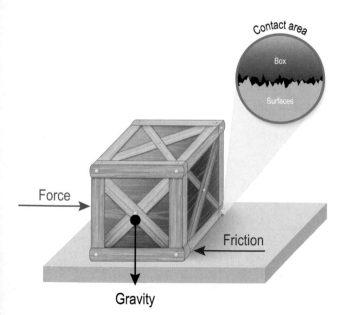

## FRICTION

**Friction** is a force that acts on an object in the opposite direction of the object's movement. It is a result of the *contact* between two solid surfaces. When you push a toy car on the floor, the force of friction acts on the car to slow it down until it eventually stops. The amount of friction between the floor and the car changes when we change the type of surface that the car rolls on. A car pushed with equal force will roll farther on a smooth, wooden or marble floor than it will on a carpet. The understanding is that the carpet has more friction than a smooth floor. This is another one of those things you probably already knew from common sense, but now you are gaining a deeper understanding of why this is.

# SURFACE VARIATION FRICTION EXPERIMENT

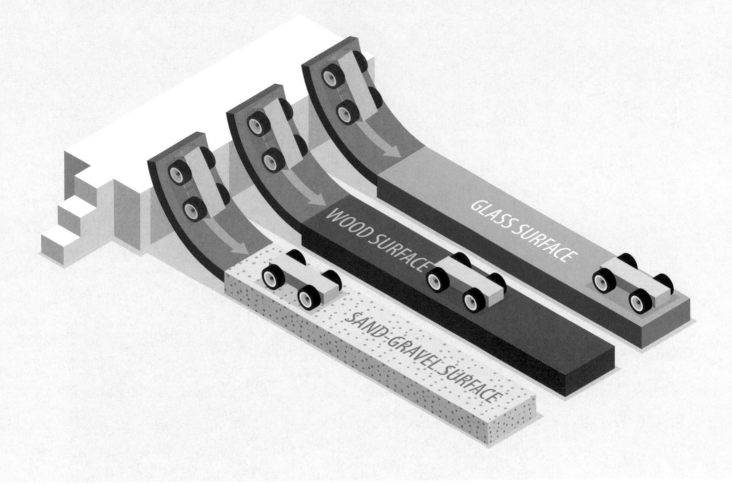

GLASS SURFACE

WOOD SURFACE

SAND-GRAVEL SURFACE

Generally speaking, a rougher surface produces more friction than a smoother surface. The reason why we need to keep pushing or pulling a toy car to continue moving it at the same speed is that we need to add enough force to surpass the amount of friction between the car and the floor.

Friction seems to be a net *negative* force since it requires us to keep pushing or pulling something to be able to move it. Yet, at the same time, friction is essential to our being able to walk or to move a car. If you have ever found yourself on a very slick surface, you probably noticed that walking becomes quite difficult. We see the same thing when streets become very icy and it is hard for a vehicle to get going or to stop. Why is this the case? Newton's third law of motion gives us our answer. As you walk, you push on the ground, and the ground pushes back on your foot. It is the reaction force of the ground that propels you forward and allows you to move. When there is reduced friction or no friction, then there would be no reaction force and you could not move forward. The slickness of the ice reduces the friction below your feet, making it harder to push off and walk (or to drive a car).

*Different surfaces—like sand, wood, and glass— provide different levels of friction. Run an experiment at home with a ball or toy car, rolling it across hardwoods, carpet, and a glass table or mirror to see this phenomenon in action.*

## DRAG

The force of **drag** is similar to friction in that it acts in the direction opposite to the motion of the object, but, unlike friction, it is caused by the fluid through which the object is moving. We will discuss fluids later in the book, but at this point it suffices to say that gases and liquids are fluids. It probably makes sense that liquids are fluids—your mom might say you need to drink plenty of fluids (water and orange juice) when you have a cold. But gases, in this scientific context, are considered fluids as well. Drag through air is called **air resistance**, or aerodynamic drag, while the force of drag through a liquid is called **hydrodynamic drag** (hydro = water). Air resistance would slow down a Frisbee on a windy day, and walking through a pool is harder and slower than walking on land because of hydrodynamic drag.

Drag is a function of the shape of the object and the density of the fluid through which it is passing. The force of drag increases as the fluid gets denser. Water is a lot denser than air, therefore it has a lot more drag than air. This is one reason why airplanes can fly faster than submarines can move. The force of drag also increases the faster the object moves. This means a car moving at 20 mph experiences a lot less drag than one moving at 70 mph.

*Drag on a parachute (as indicated by the arrows above) is what allows a skydiver to float through the sky to a soft landing!*

The shape of the object is also very important in understanding the amount of drag the object experiences. The more exposed surface area an object has, the higher the drag it experiences. You can observe this when you move your hand through water when you are at the pool. If your hand is turned sideways and made narrow, you can cut through the water like a knife (less drag), but it's much more difficult if your palm is held out flat (more drag). Carmakers do a lot of research to optimize the shape of a car to decrease the amount of drag, as less drag helps the car have better aerodynamics, and, as a result, use less fuel.

In our discussion about friction and drag, we have covered the three states of matter which objects move through, against, or across. Friction deals with objects moving against something *solid*, and drag deals with objects moving through *liquid* or *gas*. But our last force deals with something that seems a little less real, at least according to our senses. Nonetheless, we can assure you that gravity is quite real!

## Reducing Drag = Victory!

A lot of research has been done to determine how to reduce the effect of drag on moving objects. Racing teams spend more money than we can probably imagine trying to reduce the drag on their cars by just a little bit, seeking an edge on the competition. This edge comes in the form of faster speed and using less fuel (using less energy to fight drag leaves more energy to drive a little farther on the same amount of fuel, which saves you time as you are stopping less to refuel). These same efforts—though in different ways—take place in other competitive activities, including swimming. As water is denser than air, the effect of hydrodynamic drag is greater than air resistance. In the first decade of the 2000s, swimsuit manufacturers developed swimsuits that greatly reduced drag. This led to a large number of swimming records being broken in 2008 and 2009. The swimming governing body, FINA, was so concerned that the swimsuits made the competition unfair, it banned the fast swimsuits from the 2012 Olympics. And, wouldn't you know it? Very few records were broken that year!

## GRAVITY

The force of **gravity**—an attraction that pulls together all objects made of matter—is what causes an apple to fall to the ground when it is ripe. This invisible force is a favorite of little kids as they learn that all objects fall to the ground, so they push food and toys off their high chairs, annoying their parents. Gravity is also what pulls you back to the Earth when you jump up (unless you are an astronaut in outer space!).

Isaac Newton described the mathematical relation of gravity as an attractive force between two masses separated by some distance. He said that the force of gravity increases as the mass of the objects increases (so gravity acts more strongly on bigger objects), but it decreases as the objects are farther apart (so gravity weakens as the distance between two objects increases). Similarly, if two objects have a gravitational pull on one another, the larger of the objects will win that battle of gravity tug-of-war. Think of it this way: We as human beings are technically objects, because we have mass, right? But in relation to Earth, we are very, very (very) small. Earth obviously has a lot of gravity since it is so large, so it pulls us down to the planet's surface. Technically, as the planet pulls us down to it, we also pull on the planet with our own gravity, but since we are so very small in relation to the mass of the Earth, there is no effect on the Earth's motion. In fact, even if all humans alive today were in the exact same spot, we would still have no effect on the Earth.

As you know, we experience gravity constantly while on Earth. Earth's force of gravity causes objects to accelerate to the center of the planet by $9.81 \frac{m/s}{s}$. Scientists call the force of gravity that we experience our **weight**, but this is a little different than how you normally think of weight. We calculate our weight in this context by multiplying our mass by the acceleration due to gravity. So a 100 kg object has a weight of: $100 \text{ kg} \times 9.81 \frac{m/s}{s} = 981 \frac{kg \times m}{s}$ . This is a little bit more complicated than stepping on a scale and reading the pounds figure! Interestingly, this unit of measure, a kilogram meter per second per second, $\frac{kg \times m}{\frac{s}{s}}$, was named a Newton after Sir Isaac Newton, and is abbreviated N, which saves us a lot of writing.

If we were on an object with less mass than the Earth (such as a smaller planet), then our weight would immediately be lighter than it is on Earth. For example, the Moon has one-sixth (⅙) the mass of the Earth, so your weight on the Moon would accordingly be one-sixth (⅙) of your weight on Earth, due to the smaller gravitational force experienced. Conversely, your weight would increase on a larger planet like Jupiter.

A curious person might ask: If we are constantly pulled to the core of the planet, then why am I not moving as I sit reading or listening to this text? That is because the chair or bed you are sitting on is giving a reaction push in the opposite direction to prevent your motion. The floor of your house is pushing up on the chair or bed (and you) to keep you from moving. Remember, there is no net change to an object's motion when the forces are balanced.

We spoke about the game of baseball in the last chapter and return to it here, because the force of gravity explains the motion of thrown objects. When you throw a baseball from the outfield to home plate, you notice that it makes an arc. Mathematicians call this shape a **parabola**. It starts to go up and forward until it reaches a high point—or apex—and then starts falling until it hits the ground or is caught by someone on the ground. The ball does not go up indefinitely because it experiences the pull of gravity from the Earth to come back down to the surface. You can change the shape of the parabola by changing the initial angle in which you throw the ball. The same thing is experienced by anything that moves through the air.

The curious reader might ask another question: If things will always fall back down to Earth, then how can we get objects into space? When we send a rocket into space, it seems like there is no gravity, right? Of course, there actually is—there is a lot of gravity pulling on a huge spacecraft. But you probably

*A space shuttle must use a lot of energy to get off the ground as it fights the force of gravity, but once it reaches outer space, it can float rather effortlessly.*

know that these rockets require an incredible amount of energy and thrust to escape Earth's gravity—perhaps you have seen a shuttle launch and the huge explosion that ignites beneath it. If there is enough power behind something, it is possible to, if you will, "cheat" gravity (just think of flying on an airplane).

If a shuttle is moving fast enough and gets high enough, it will enter the weightlessness of space, where there is much less gravity (although not none, as some people mistakenly believe). The shuttle will enter an orbit where it is, in a sense, constantly falling and constantly missing Earth. Astronauts in orbit experience the state of weightlessness because they are constantly falling, not because Earth has ceased to have an effect on them. We escape orbit and send astronauts to the Moon by adding extra force to "jump" out of the fall. We must add this extra force, this extra acceleration, to counter the attraction from Earth and to move in the direction of a different object. It will work the other way too. When the first astronauts reached the Moon in July of 1969, they entered the Moon's orbit and, when they wanted to return home, had to accelerate to leave that orbit and come back to Earth.

With our study of these forces behind us, we will move on in the next chapter to discuss another important phenomenon that permeates the world all around us—energy!

## FOUNDATIONS REVIEW

✓ Friction is a force that acts on an object in the opposite direction of the object's movement. It is a result of the contact between two solid surfaces. Generally speaking, a rougher surface produces more friction than a smoother surface. Friction is essential to our being able to walk or move a car. As you walk, you push on the ground, and the ground pushes back on your foot. It is the reaction force of the ground that propels you forward and allows you to move.

✓ The force of drag is similar to friction in that it acts in the direction opposite to the motion of the object, but, unlike friction, it is caused by the fluid (gas or liquid) through which the object is moving. Drag through air is called air resistance, or aerodynamic drag, while the force of drag through a liquid is called hydrodynamic drag (hydro = water).

✓ The force of gravity is an attraction that pulls together all objects made of matter. It is what pulls you back down to Earth when you jump up. Isaac Newton described the mathematical relation of gravity as an attractive force between two masses separated by some distance. The larger the mass, the stronger the pull of gravity on it.

# Spiritual Forces

In this book we examine some types of forces that can be applied to objects. Physicists and engineers have become very good at using controlled forces to hold buildings in place or move objects in a given direction. For example, we can use the forces acting on a falling object in orbit around a planet to accelerate it and "slingshot it" out of orbit toward a moon or another planet in the solar system. The object to which the force is applied has no option but to obey the force, as it has no "agency," meaning no will of its own, or no ability to make its own decisions.

We humans do have agency because we have free will, which allows us to make decisions. And we, too, face forces. Yes, physical forces like gravity act on us, but also spiritual forces in the sense of temptations and our disordered passions. Unlike inert objects, or even animals, we can choose whether to do the right or the wrong thing because our rational nature includes the will. The object of the will is to choose the good.

Before Adam and Eve disobeyed God, they had the gift of integrity. This gift made their passions subject to their will, which was subject to the intellect. The serpent did not tempt Adam and Eve's passions, he tricked them by appealing to their intellectual knowledge of the good. Being like God is what we were made for, which is why the souls in heaven participate in the life of the Trinity. Satan tricked Adam and Eve by telling them that God was keeping a good from them that could be attained without God. When humanity reached after divinity instead of receiving God's gifts, we lost our friendship with God and the preternatural gifts we had been given.

We children of Adam and Eve live with the loss of their original innocence. Our intellect is clouded and our passions are disordered. We are now assailed by these passions, which tempt us to do that which keeps us from God. We are assailed by the world, which offers partial goods that may keep us from our eternal good. However, we are capable, with God's grace, of resisting these temptations and still achieving our eternal end. God's grace is the spiritual force applied to our souls to keep us on the path to heaven.

Gas turbine engines convert natural gas (and other liquid fuels) into mechanical energy, which is eventually transferred (via a generator) into the electrical energy that powers our homes.

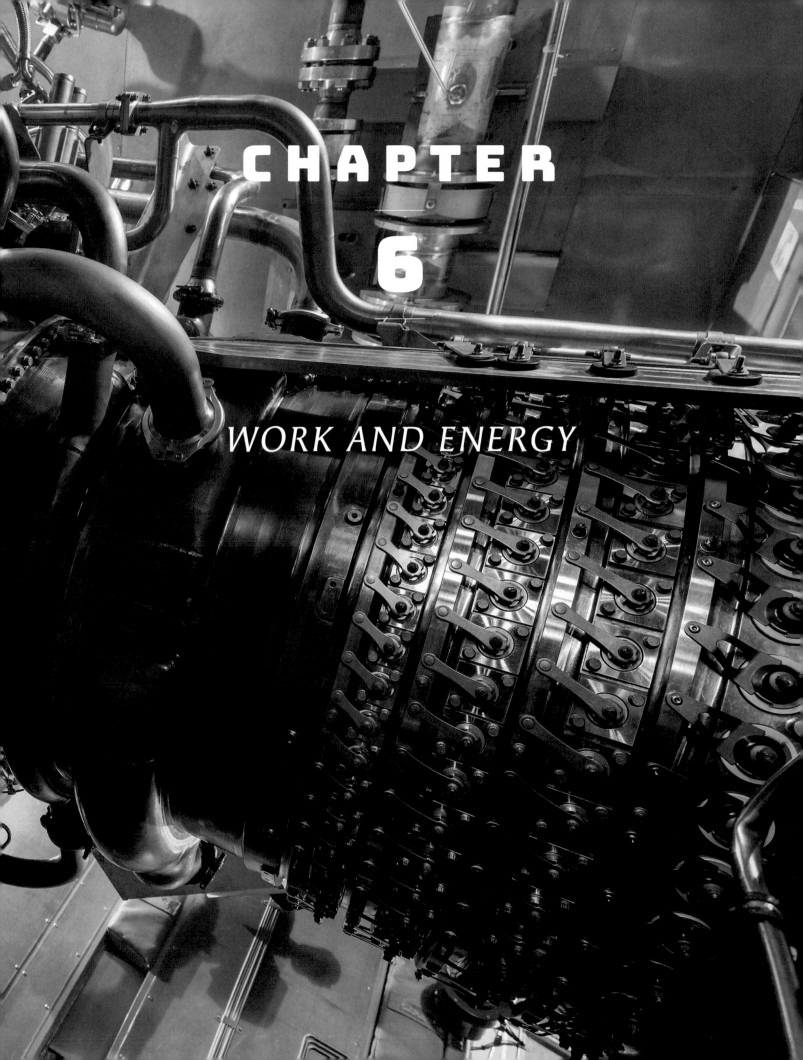

# CHAPTER

## 6

*WORK AND ENERGY*

## ENERGY IN EVERYDAY LIFE

After looking at how forces are used to cause and change motion in the last chapter, we can now look at another important concept in the scientific world—energy! The idea of energy is a key to understanding the way we live in the age of lights, computers, and machines. One day you will likely pay a power bill (your parents pick up that tab now). In essence, this is a payment for the energy your household uses in the form of electricity. Similarly, you know that most cars need to be filled with gasoline for their engines to work. Gasoline is the energy source for the car. These are just a few common examples of how we depend on energy in everyday life. But there is a lot more to it. Let's start off our journey into the world of energy by looking at Newton's mechanics.

## UNDERSTANDING WORK

Work is one of those words that has a specific definition in science that is different from the way it is commonly used. We speak of going to work, or doing homework, or working in the yard to rake up leaves. We basically mean we are expending energy—physical or mental, or both at once—to accomplish some goal. But the scientific definition is a little different. **Work** is the transfer of energy when a force acts within a certain distance. This means that mechanical (movement) work is done when a force is used to pull or push (or lift) an object a given distance. Before we get to some examples to help clarify what this means, let's look at a mathematical equation that demonstrates what work is.

Below you will see work represented by "$W$." "$F$" is the force in Newtons, and "$d$" is the distance in meters.

$$W = F \times d$$

The unit for work is given as the Joule, represented by J. Joules is a measure of work, in the same way that inches can measure your foot. There was an English physicist named James Prescott Joule who made great contributions in this area of science, so work units are named after him.

Now that we have that mathematical foundation before us, let's look at a few examples.

Let's say a 100 N force is used to move an object 10 meters. Maybe you want to push that toy car 10 meters and that is the force at which you are pushing it. We find the work as follows:

$$W = 100 \text{ N} \times 10 \text{ M} = 1{,}000 \text{ J}$$

Or let's say you are not pushing an object forward but rather lifting one up from the floor to a counter. In this case we would use the weight of the object (remember that force equals mass times acceleration) and the height it was

*The unit for work is named for James Prescott Joule (1818–1889).*

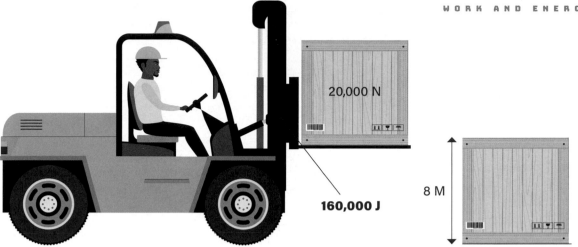

20,000 N

160,000 J

8 M

lifted to find the amount of work. For example, let's say a forklift uses a force of 20,000 N to lift a crate onto a storage rack 8 meters above the ground. We would calculate the work done by the forklift in a similar way:

$$W = 20{,}000 \text{ N} \times 8 \text{ M} = 160{,}000 \text{ J}$$

Notice that in both examples the work is done by the action of a force acting on an object. We would say that if 1,000 J of work are done pushing an object across the floor, then the person pushing the object across the floor used 1,000 J of energy in pushing the object. Similarly, as the forklift raised the object and did 160,000 J of work, then it had to spend 160,000 J of energy in doing that work.

Some of this may be hard to get your mind around. It's easy to think of measurements when you want to know how many inches your foot is or how many pounds you weigh. But the concepts of force exerted and the amount of work done to lift something—scientifically speaking—are harder to understand and express in plain terms. Now you can see why we use the word "work" the way we do in common language. You don't take time to calculate the work in mathematical or scientific terms each time you lift or push something—you just know you are *working* to do it. You are expending energy.

Speaking of energy, both these examples we just discussed point to the scientific definition of energy. In everyday language we say we need energy to go do something, such as to work or play. Or, we might say batteries give a toy energy so that it works. You can see how energy and work are related. **Energy** is the capacity ("ability") to do work.

Notice in our equations that the amount of work done and the amount of energy spent balance each other. This is because energy is one of those physical quantities that is conserved. Scientists call this the **law of conservation of energy**, which states that energy is neither created nor destroyed, only converted or transferred from one kind of energy to another.

But where does the energy originate from? Where did the energy used by the person come from? Where did the energy in the forklift come from? To answer those questions, we first must understand a little more about energy.

*The forklift is doing the work necessary to lift the crate. But it needed energy in order to preform this work. Where did that energy come from? We'll learn about that on the next page!*

## Entropy: An Unfortunate Reality

A key idea in physics is that of the conservation of energy. Even if we had an apparently perfect system with no obvious losses of energy, such as that lost to friction or drag, energy is always lost or wasted somewhere and somehow. Scientists came up with a measurement to account for this, which they call entropy. In mechanics, which we have defined as the study of the movement of objects through space, **entropy** is used as a measure of wasted heat. Since entropy is impossible to avoid and is present everywhere, scientists coined a law of thermodynamics which states that entropy is always increasing in the universe. This means that there is always less energy available to do work than the amount of energy we have. For example, the gasoline in your parent's car has a given quantity of potential energy, but, due to entropy, the total energy available to do this work is lower than the potential energy stored in the gasoline (this basically means that even though there may be twenty gallons of fuel in the tank, you won't get twenty gallons worth of energy out of it because of entropy). When we look at our understanding of the structure of material things, we will see how scientists understand the source of entropy in a process.

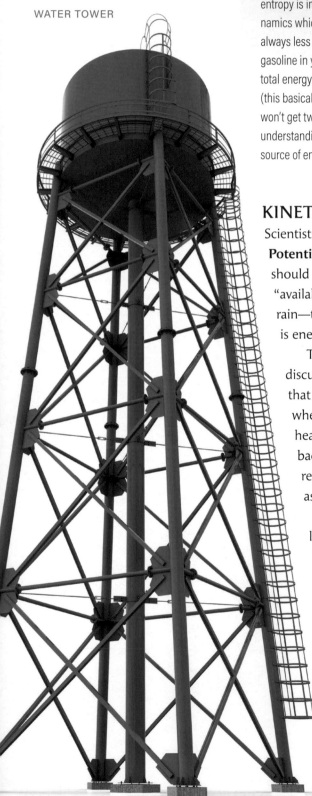

WATER TOWER

## KINETIC AND POTENTIAL ENERGY

Scientists speak of two types of energy: potential energy and kinetic energy. **Potential energy** is energy that is stored up to be able to do work. It should make sense to us that potential energy is energy that is stored or "available." When you see a storm cloud rolling in, there is a *potential* for rain—there is rain *stored up within the cloud*. With potential energy, there is energy stored up in something.

To understand this, let's return to the same toy car we've been discussing so much. This time imagine it's not an ordinary toy car that you just push forward, but rather one that has a pullback motor, where you have to drag or *pull it back*, and you feel the tension and hear the click of the clockwork motor being wound up. By dragging it back, you have *stored up potential energy*. You need only to let it go to release this energy and watch it race across the kitchen floor (which as we'll see in a moment is kinetic energy at work).

Another example of potential energy is that forklift that was lifting a load. If it took the forklift 160,000 J of energy to lift the heavy load, we can say that it now has stored 160,000 J of energy that can be "used." If you think about it, the higher you go to drop a rock, the faster you would expect it to be moving as it hits the ground. The *higher* an object is, the *higher* the amount of potential energy that can be converted to kinetic energy. A great example of this is water stored in a water tower that needs to be pumped to the town. The water is stored at an elevation because that height can be used to turn the water into moving water, to eventually provide water service to the community.

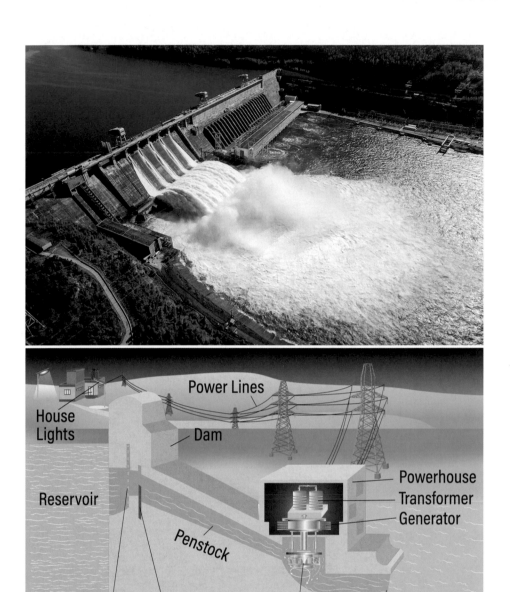

Labels on diagram: Power Lines, House Lights, Dam, Reservoir, Penstock, Powerhouse, Transformer, Generator, Intake, Control Gate, Turbine, Outflow

*Hydroelectric dams (above) and hydropower plants (below) provide energy by turning a column of water into an electrical current.*

Another example is that of a hydroelectric dam, which turns a column of water into an electric current. Other types of potential energy that we will discuss later in the text include electrical potential and chemical potential energies, such as those found in fuels.

Now let's consider kinetic energy in more detail. With our favorite toy car, as we said, once you have wound the motor back (storing up all that potential energy) and released it, the energy of it moving forward is kinetic energy. **Kinetic energy,** then, is the energy of objects that are in motion and/ or doing work.

There are different types of kinetic energy, but at this point we will concentrate on the kinetic energy based on Newton's mechanics and work. All moving objects have kinetic energy that *increases* as the speed *increases* and *decreases* as the speed *decreases*, with a minimum of zero once the object slows to a stop.

An object that gains speed and kinetic energy has work done on it, and the energy must come from a source. For example, as your car accelerates from a red light to reach the speed limit, the kinetic energy increases by using the chemical potential energy stored in the gasoline, diesel, or batteries.

Recall our law of conservation of energy, which taught us that energy doesn't disappear, it gets converted into different energy. Well, then, where does the energy of the car go when the car comes to a stop at the red light? Here is where friction comes into play, which we spoke about a while back. Friction from the brakes against the road heats up the brakes. So, in this case there is a change of one type of kinetic energy—movement—to another: heat.

## HEAT

Heat is a type of kinetic energy that comes about from differences in the temperatures of things. Heat always moves from a hotter object that has a higher temperature to the cooler object that has a lower temperature—in other words, heat transfer is a one-way street. The amount of heat that moves is calculated by using the temperature difference between the objects, the mass of one of the objects, and what is called the heat capacity of the substance. The **heat capacity** of a substance is how easily some mass of the substance changes temperature—how quickly it heats up or cools down. Some substances take very little energy to change temperature, like metals that cool or heat up easily. On the other hand, materials like Styrofoam cups take more heat to change their temperatures. Materials that readily get hot or cold are called **conductors**. They are used when heat needs to be moved; for example, we use metal to make cooking pans because we want the heat from the stove to readily transfer to

*Physics Fun Fact:*
*Water can be heated up and cooled off. It only stops changing temperature when it's temperature is equal to the room temperature in which it is found.*

*We use metal pots and pans to cook our food because they make good conductors, meaning they transfer the energy from your stove to the food being cooked.*

the food to cook it. Materials that do not easily change temperature are called **insulators** and are used when we need to keep heat in or out of somewhere. For example, if you need coffee to stay hot for a long time, you place it in an *insulated* container which prevents the transfer of heat. The insulation in the container would also protect your hand from getting burnt while you hold it, whereas with a hot pan (a conductor) you would need an oven mitt.

As we stated, the quantity of material is also important in determining the amount of heat transfer needed to change the temperature of a given substance. You can easily infer this by noting that if you have one cup of ice-cold water and place it next to a gallon jug of ice-cold water, the gallon jug will stay cold a lot longer than the cup of water—this is because it holds a lot more water. The extra mass (the extra water) helps keep it cool longer.

One might wonder when the water will stop changing temperature. That will happen when the water reaches the same temperature as the room in which it is placed, as the transfer of heat requires temperature *differences*. When the water is at the same temperature as the room, it will stop changing temperature, and we can say that it and the room have reached a "state of balance," or **thermal equilibrium**. This means no more heat can be moved.

In our next chapter, we will continue our discussion of energy by pivoting to discuss light and sound waves, which allow us to see and hear things.

## FOUNDATIONS REVIEW

✓ Work is the transfer of energy that takes place when a force acts within a certain distance. This means that mechanical (movement) work is done when a force is used to pull or push an object a given distance.

✓ Energy is the capacity ("ability") to do work. The amount of work done and the amount of energy spent balance each other. There are two types of energy: potential and kinetic. Potential energy is energy that is stored up to be able to do work, while kinetic energy is the energy of objects that are in motion and/or doing work.

✓ Heat is a type of kinetic energy that comes about from differences in the temperatures of things. Heat always moves from a hotter object that has a higher temperature to the cooler object that has a lower temperature. The amount of heat that moves is calculated by using the temperature difference between the objects, the mass of one of the objects, and what is called the heat capacity of the substance. The heat capacity of a substance is how easily some mass of the substance changes temperature.

# Finite Versus the Infinite

In the material world, the quantity of energy stored in a system is finite (there is a set limit) and can only be used to do one certain useful thing at a given time. Not only that, the energy available to do work is always lower than expected due to entropy losses, as we just learned. While the world's engineers have improved ways of reducing the wasted heat produced due to entropy, our energy supply is still limited. This limit exists because material things are limited. After all, you can only put so much fuel in the ten-gallon gas tank of your car, right? This is analogous to having ten pieces of candy to pass on to your friends. If you give all the candy to one of your friends, then no one else gets a piece. Alternatively, if you have five friends to give candy to, then each of them can get two pieces but then the candy runs out.

Nonmaterial things do not share these limits. You do not have a finite store of love you can pass around to your friends and family. You can fully love each of them without your love "running out." And, in a very interesting way, the more you love people, *the higher your capacity to love becomes*. This applies to other nonmaterial things such as

justice, happiness, and mercy. Think of the happiest day of your life. Even that happiness pales with the happiness that is available to you if you achieve the union with God that is offered to you in heaven.

Just as nonmaterial energies are limitless for you, their abundance and infinity are even more apparent in God. God in Himself lacks nothing. His act of creation is a free act of sharing His love with us. In other words, the act of creation is not necessary for God to fulfill His Being; it does not add any glory to God, nor take anything away from Him. Similarly, the graces earned by our Lord in His Passion are infinite, and there is nothing lacking in them. They completely and efficaciously gain salvation for all men.

If His salvific act is perfect and complete, does this mean all men will achieve the beatific vision? And if not, why do we say the work is "complete"? Remember that each of us has a will, and we can each freely choose to reject the graces gained by Christ. No one can force us to be good Catholics. In a mysterious way, through our free will, we must choose to become and remain partakers of the divine nature.

*Though we cannot see them with the naked eye, sound waves, shown here in vibrant colors, travel through the air and allow us to hear things.*

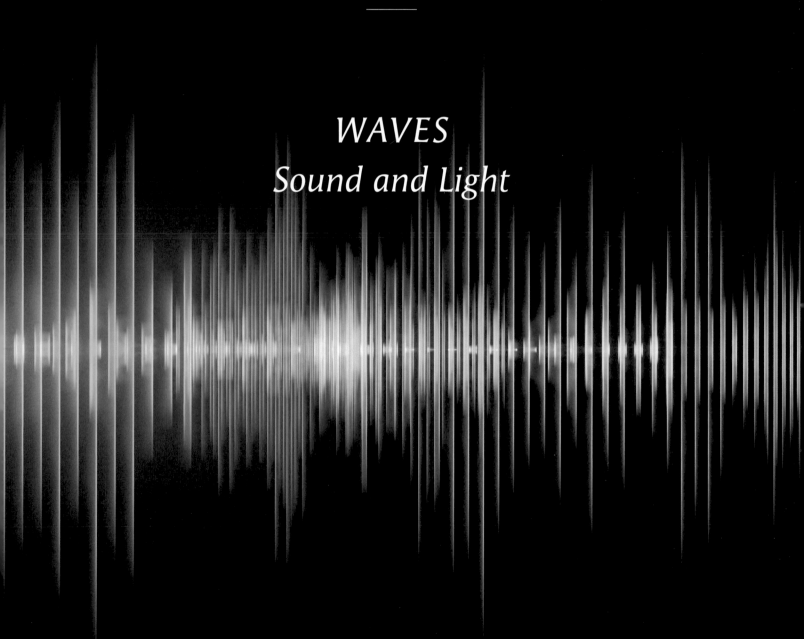

# CHAPTER

## 7

*WAVES*
*Sound and Light*

## WHAT IS A WAVE?

You have most likely dropped a rock into a pond or lake before and heard the "plop" sound that it makes upon landing. And if you have, you know that concentric circles ripple out from the spot where the rock hit the water. It happens very quickly, but if we could slow down time and zoom in, we would see that the ripples have peaks where the water is higher and valleys where it is lower, like a circular mountain range. Both the sound you hear and the ripples you see moving away from the point of impact are ways in which the kinetic energy of the rock is *dissipated* in both the air and the water—meaning the energy is scattered out. We say that the energy of the rock is dissipated by the repeating pattern of motion we observe, and this is called a wave.

*Physics Fun Fact:*

*The average guitar can play dozens of different kinds of notes, each made by strums on the strings (vibrating them) combined with the unique arrangement of the musician's fingers on the guitar's neck and fingerboards (as they hold down the strings).*

**A wave** is a transfer of energy created by a disturbance. In the example above, the *disturbance* is the rock hitting the water. Similarly, as you tap your fingers on the table, the tapping sound you hear is caused by the disturbance of your fingers hitting the surface of the table. If you pluck a guitar string, the disturbance causes the string to move, which causes the air around it to move, which causes the pleasant sound you hear as a musical note.

Waves can be characterized in two main ways. The first way to differentiate them depends on the medium in which they are transmitted, such as through water or air or a solid (or through no matter at all)—meaning we tell them apart based on *what the wave is moving through*. The second way to differentiate waves is related to the pattern in which they move—how they move and in what direction.

Let's discuss these two ways we classify waves, each having further breakdowns within them. We will start with the medium in which they are transmitted.

## MECHANICAL AND ELECTROMAGNETIC WAVES

All of the examples we have listed so far (the rock hitting the water, tapping your finger on the table, plucking a guitar string) are examples of mechanical waves. **Mechanical waves** are those that transfer energy by *moving matter* (while moving through matter) in an oscillating or repeating pattern. Think back to the pond water rippling in circles after being struck by the rock. As we already mentioned, you see the water make a repeating pattern of peaks and valleys. The particles of water move up and down in place (the ripples extend out), transferring the energy away from where the rock fell. Once the energy is completely dissipated the water goes back to its normal state. This is a mechanical wave because it is moving through matter (in this case, water).

All sound waves are mechanical waves—think of the guitar string being plucked. It is also moving through matter (air). Even things you may not think of as waves, such as earthquakes, are mechanical waves. The energy that causes the quake—the disturbance of tectonic plates grinding against one another—is dissipated through the Earth's crust in what are called **seismic waves**.

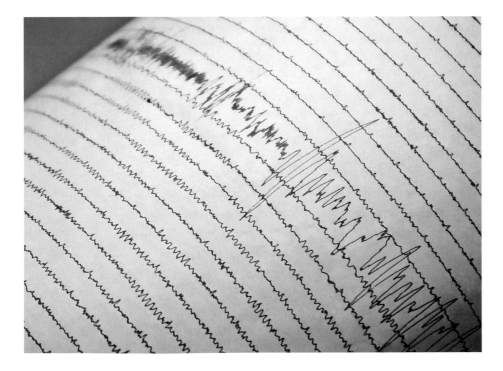

*The power of an earthquake can be recorded on a seismograph, which measures the size of the seismic waves traveling through the earth's crust.*

Mechanical waves move at different speeds and, generally speaking, move faster through denser things, such as solid objects.

The second type of waves are **electromagnetic waves**. These are considered to be pure transfers of energy that require *no matter* to move the energy (though they can move through matter as well). As such, they move fastest when there is no matter. This means that electromagnetic waves move fastest in a **vacuum**—a region of space with little or no matter. The electromagnetic wave you are most familiar with is the one that allows you to read this text: light! Microwaves, radio waves, UV-rays, and X-rays are all examples of electromagnetic waves. Unlike mechanical waves, all electromagnetic waves move at the same speed when in a vacuum.

Therefore, the simplest way to distinguish these two waves is that electromagnetic waves do *not* need a medium of matter to travel through, while mechanical waves do (whether it be air or gas, water or other liquid, or a solid surface).

## TRANSVERSE AND LONGITUDINAL WAVES

Our second way of characterizing waves is by *how* they dissipate the energy in relation to the point of disturbance—essentially in what direction. To understand this, think of you and a friend holding different ends of a stretched-out slinky, or string. When you start moving your hand up and down, while your partner does nothing, you see the waves moving *toward* your friend. The motion of the wave is perpendicular, or at a right angle, to how you are moving the string (you are rippling the string up and down and the wave is moving out "sideways" from you). This type of wave is called a **transverse wave**. The motion of the ripples of water from the point in which the rock hit the surface is another example of a transverse wave as the rock was moving *straight down*, and the waves are moving *out* (sideways) from the point of contact.

The second type of wave is easiest to see with a slinky. In this case, instead of moving your hand up and down, you would push and pull the slinky in the same direction that it is being held. In doing this, you observe that there is a section in which the coils of the slinky are *close to each other* and another in which they are *far apart*. If you were to concentrate on the compressed part in which the coils are close to each other, you would notice that the wave is moving in the same direction in which you are pushing it—away from you. This type of wave, in which the wave travels in the same direction as the disturbance (they travel parallel to each other), is called a **longitudinal wave**. Waves of the ocean and sound waves are longitudinal waves. With our slinky example, the part where the coils are farther apart is called the *rarefaction*, and where they are close together is called the *compression*.

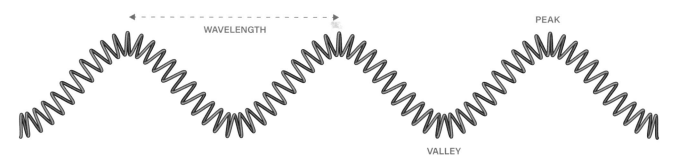

**TRANSVERSE WAVE**

WAVELENGTH

PEAK

VALLEY

WAVE DIRECTION

**LONGITUDINAL WAVE**

COMPRESSION          RAREFACTION          COMPRESSION

## WAVE SPEED, WAVELENGTH, AND FREQUENCY

We can also describe waves by giving their speed, the length of the wave, and how many waves are measured in a given time. These three features are related to each other.

*Wave speed, frequency, and length all have a relation to one another.*

**Wave speed** is similar to the speed of anything else—it's how much distance the wave covers in a given unit of time (second/minute/hour). Generally speaking, we measure wave speeds in meters per second ($^m/_s$). Sound moves on average at about 340 $^m/_s$ through the air—this means it's traveling just over 1,000 feet in a second! If you think that's fast, though, light travels in a vacuum at an astonishing 300,000,000 $^m/_s$—that's almost a million times faster than sound! Since light is so much faster than sound, this explains why during a thunderstorm you see the lightning before you hear the accompanying thunder boom. Or, if you are watching a golfer strike a golf ball from far away, you will see him swing first and hear the pop of the club against the ball a second or two later.

Wavelength and frequency are related to the speed. The **wavelength** is the distance between repeated parts of the oscillating pattern. For a transverse wave, scientists measure either from peak to peak or from valley to valley. For a longitudinal wave, they measure from the beginning of a compression to the beginning of the next compression. Meanwhile, the **frequency** is how many waves are measured in a specific point in space for a given unit of time. Say you are at the beach and count six waves in a minute. Then the frequency of waves would be six per minute. The standard timeframe for counting waves is a second, so the frequency is given in waves per second, which is called a Hertz (Hz).

We noted how wavelength and frequency are related to speed. The following equation shows how:

$$\text{Speed} = \text{Wavelength} \times \text{Frequency}$$

So we calculate the wave speed by multiplying its wavelength (the distance) by the frequency (the time). This means wavelength and frequency are *inversely proportional* to each other, meaning that at constant speed, as the wavelength increases, the frequency decreases, and vice versa. Let's plug in some numbers to show this:

$$20 = 10 \times 2$$
$$20 = 5 \times 4$$

Do you see how, when the speed holds at 20, if the wavelength is cut in half (from 10 to 5), then the frequency is doubled (from 2 to 4). It works the other way, too:

$$20 = 10 \times 2$$
$$20 = 20 \times 1$$

In this example, the wavelength was doubled (from 10 to 20), so the frequency was cut in half (from 2 to 1). This is what it means when something is inversely proportional—as one moves up, the other moves down. In terms of waves, this means that as the wavelength stretches out, the frequency of the waves will go down. But if the wavelength shortens, the frequency increases. This makes sense, doesn't it? Think if you had long blocks and short blocks to fit inside a box. The box would fill up with *fewer* long blocks (longer wavelength = less frequency) and would need *more* short blocks to fill up (shorter wavelength = higher frequency).

Remember that all electromagnetic waves move at the same speed in a vacuum. This means that what differentiates electromagnetic waves are the frequency and wavelength of the waves.

## SOUND WAVES

As mentioned earlier in this chapter, sound waves are mechanical, longitudinal waves. They move through the air at about 340 $^m/_s$. If sound waves move at the same speed, why doesn't everything sound the same? Well, because what makes the different pitch of the sounds you hear are not differences in wave speeds but in the frequency of the waves. Your ear is capable of hearing sound waves between 20 Hz and 20,000 Hz. At lower frequencies you hear low-pitched sounds—think of the keys at the far end of a piano that make

*Physics Fun Fact:*

*Sound moves on average at about 340 $^m/_s$ through the air—this means it's traveling just over 1,000 feet in a second! If you think that's fast, though, light travels in a vacuum at an astonishing 300,000,000 $^m/_s$ — that's almost a million times faster than sound!*

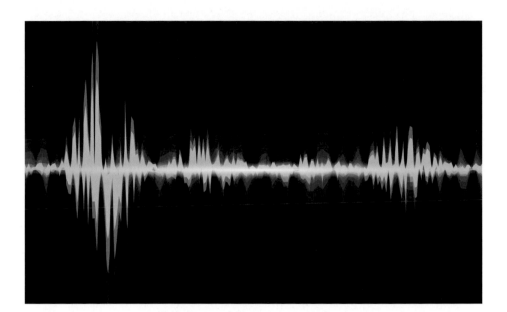

*Can you tell where the low-pitched frequency versus the high-pitched frequency would be by looking at these sound waves?*

an ominous thunder-like *boom*. As the frequency increases, the pitch of the sound increases—you will hear this in the higher notes on the opposite end of the piano, notes that sound almost like a teakettle whistling. This might make sense, if you think about it. Lower-pitched sounds have a kind of slow, methodical drone to them (low frequency), while high-pitched sounds have a kind of frantic, frenzied sound to them (higher frequency). This may be a good way to remember it.

*The keys of a grand piano are always positioned the exact same way, with the low frequency keys on the far left, and the high on the far right. With the keys in the same position this allows pianists to sit down at any piano and know how to play.*

*Echo Canyon near the Chama River in New Mexico acts as a natural amphitheater. Echoes bounce off hard surfaces like granite and rock, while softer surfaces absorb the sound.*

Usually, though, we think of a sound not in terms of pitch but in terms of how loud it is. The loudness of the sound is related to how much energy is being moved by the wave. If you barely tap the table, you will hear a very faint sound. If you hit the table harder, then the sound is louder since you gave it more energy. When you increase the volume of your sound system, what you are doing is increasing the energy used by the system to make sound waves.

We can use sound waves to understand a couple of important properties of waves. The first is that waves can bounce back when they hit a surface. This is called **reflection**—think of a mirror "bouncing back" your reflection. The main example of this in sound would be that sound echoes. For sound to echo, the sound waves must hit a hard surface so that they bounce back. Softer surfaces tend to absorb the sound, so that there is no echo. If you were to yell in a large, empty gymnasium, you may hear the sound of your voice echoing back. But, if you are in a movie theater with soft carpet and hundreds of comfortable chairs, the noise gets absorbed and does not echo back.

The second property of waves is that of resonance. This one is important in the making of musical instruments. **Resonance** refers to when waves of the same frequency vibrate with each other and their effects get amplified, or added to one another, and they sound louder. If you were to remove a guitar string and pluck it in the middle of an empty room, then it would vibrate in place and make only a faint sound. But once you place the string in a guitar's body and pluck it, the sound is a lot louder, as the waves in the cavity of the guitar (the inside of the wooden part) add their effects and resonate with each other to give a louder sound. When you strike a piano key, it causes the strings to vibrate. The notes resonate on the actual wood of the soundboard to make a louder sound.

*The differing shapes of theses stringed instruments create different sounds when musicians play them.*

## LIGHT WAVES

Just as sound moves in waves, so does light. Light waves are electromagnetic waves with wavelengths between 400 nm and 700 nm (nm = nanometer, one billionth of a meter). This is the whole range of light that is visible to the human eye—anything less or anything more and we cannot see it (unless we have some instrument to help us). Different colors have different wavelengths, but when the whole spectrum of light is seen together, we see white light. You see light reflected off a mirror, and since we are so used to mirrors, it helps us see that property of waves.

### LIGHT WAVES VISIBLE TO THE HUMAN EYE

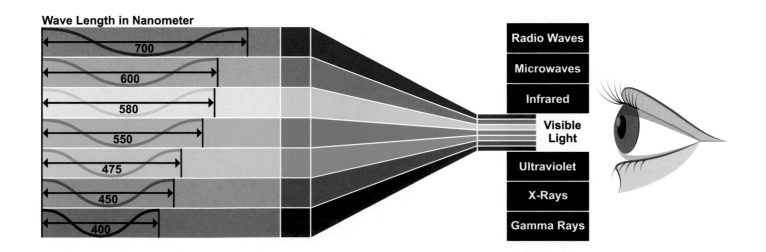

**Wave Length in Nanometer**

700
600
580
550
475
450
400

Radio Waves

Microwaves

Infrared

**Visible Light**

Ultraviolet

X-Rays

Gamma Rays

*A prism has refracting surfaces at sharp angles that separate white light into a beautiful spectrum of colors.*

Light can also be reflected imperfectly, meaning some light is reflected by a surface while other light is absorbed. Any time you see a color other than white or black, you are seeing the colored light that is reflected by that surface, while the other light is being absorbed. For example, the green grass at the park is reflecting green light to your eyes and absorbing the other light colors. You might think of it as the grass is taking in and "hiding" the other colors from you (absorbing them), which leaves only the green light for it to show. In a plant like grass, the energy absorbed by the other light colors is used for photosynthesis. A white-colored object is one that is reflecting the entire spectrum of light, while a black-colored surface is absorbing the whole spectrum of light. This is why it is better to wear white in the summer when the sun is intense—it reflects more of the light so that your body does not absorb as much heat/energy.

The last property of waves that is important and is most easily seen with light is **refraction**—the bending or redirection of light as it passes from one medium to another. Remember, light moves faster in a vacuum and slows down in denser mediums. So, it would move faster in a vacuum than in air, and faster in air than in water. When light travels from one medium to another, its speed changes and the path of the light *bends in the direction of the denser medium.* You can see the effect of refraction if you look at a straw or spoon in a glass of water. From the side, it may look like the straw or spoon is broken or bent. Refraction is studied using glass or plastic prisms that can be used to bend light. If light is bent enough then it disperses into the different colors of the light spectrum, making a rainbow. This is because the amount of bend is slightly different for the different colors of the spectrum—different degrees of bend reveal different color shades of light.

With our discussion of waves behind us, let us move on in our next chapter to a scientific discovery that changed the world—electricity!

## Light and Grace

One of the most impressive things about European cathedrals (and some here in the United States) is the stained-glass depictions of events in the life of Our Lord, Our Lady, or the lives of the saints. As the light of the sun hits the stained glass, different parts of the spectrum of light are absorbed by the various pieces of glass, taking the white light of the sun and "painting it" with color. This is analogous to how we can see God's grace in the lives of the saints more vibrantly than in the rest of us. We see the effect of God's grace most clearly when we study the lives of the great saints, who show what is possible when we live in a way that participates in the life of the Trinity.

## FOUNDATIONS REVIEW

✓ Waves are transfers of energy created by a disturbance. They can be characterized in different ways, including the medium through which they travel (mechanical versus electromagnetic) and the pattern/direction in which they move (transverse and longitudinal).

✓ Waves can be measured by speed, length, and frequency, all of which are related to one another. Wave speed can be calculated by multiplying the wavelength by the frequency, which means that at a constant speed wavelength and frequency are inversely proportional. The different pitch sounds we hear are not differences in wave speeds but in the frequency of the waves. The human ear is capable of hearing sound waves between 20 Hz and 20,000 Hz. At lower frequencies you hear low-pitched sounds, and at higher frequencies you hear high-pitched sounds.

✓ Light waves are electromagnetic waves with wavelengths between 400 nm and 700 nm. This is the whole range of light that is visible to the human eye—anything less or anything more and we cannot see it unless we have some instrument to help us. Different colors have different wavelengths, but when the whole spectrum of light is seen together, we see white light. If we are seeing a certain color—like green—then that means all the other colors are being absorbed and green is being reflected.

# Sacred Music

The Catholic Church has declared the pipe organ as the musical instrument of choice for music played in a liturgical setting. Pope Saint Pius X wrote a papal document called *Tra le Sollecitudini* explaining the role of sacred music in the liturgical life of the Church. In it he stated that liturgical music is primarily vocal, but the organ is the primary musical instrument that may be used along with human voices in the liturgical functions of the Church.

The pipe organ works similarly to the blowing of air in a glass bottle partially filled with water. As the level of the water drops and the height of the air in the bottle increases, the pitch lowers. The pitch of a note we hear is directly related to the frequency of the sound wave making the sound: the higher the frequency, the higher the pitch of the sound. Also, remember that if the speed of a wave is held to be constant, as the speed of sound inside a church would be, each wavelength corresponds to a particular frequency.

A pipe organ has a multitude of pipes through which air is blown. The organist is able to select which pipes will have air go through them. He may select one set of pipes or multiple sets of pipes to play together. The more pipes that play the same note, the louder the sound will be, as the effect is additive.

The sound of a note is controlled by the shape and material of the pipe. The ratio of the diameter to the height of the pipe determines the sound wave that will resonate in the pipe. Thus, to change the sound made by a given pipe, the organ maker would make the pipe of just the right size. Pipes of the same material and dimensions would make identical sounds. The material of the pipe changes the sound because each material has different natural vibrations that resonate differently when the note is played. This is also the reason why the same note sounds differently when played by a trumpet (brass), flute (woodwind), violin (string), etc. Different metals and different woods have different natural resonances and they all add to the amazing instrument that is a pipe organ—truly one of the great gifts of the Catholic Church to the world.

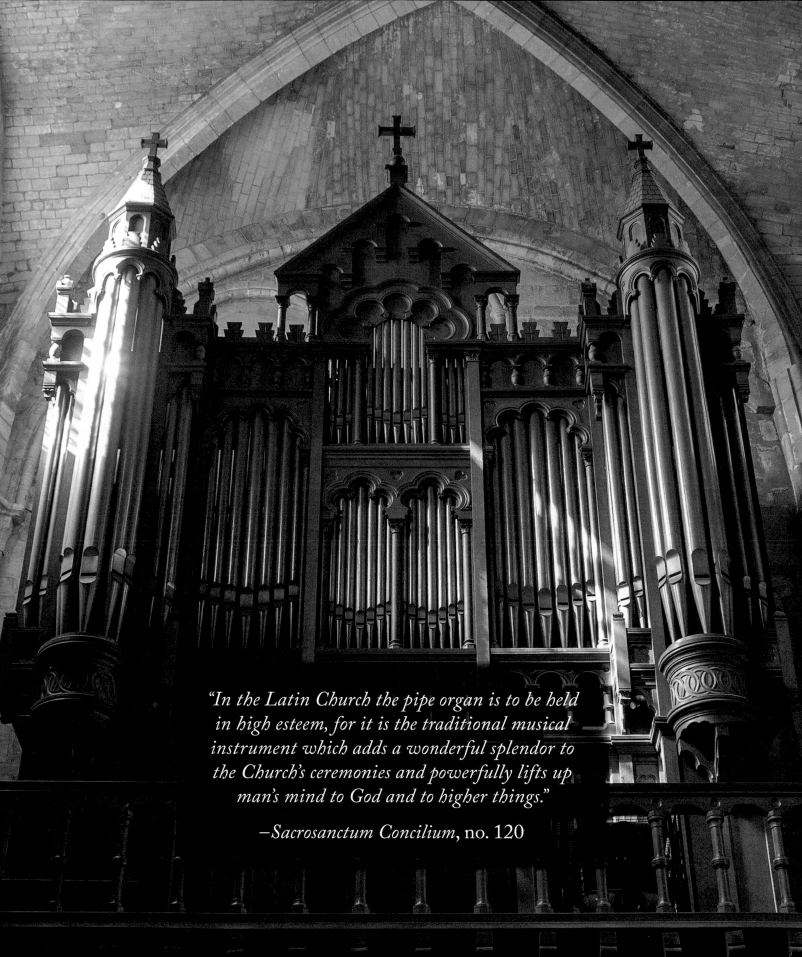

"In the Latin Church the pipe organ is to be held in high esteem, for it is the traditional musical instrument which adds a wonderful splendor to the Church's ceremonies and powerfully lifts up man's mind to God and to higher things."

—*Sacrosanctum Concilium*, no. 120

Electromagnetism—the science of charge and the forces and fields associated with it—has changed the world arguably more than any other scientific area of study. Spark coils like this one are found in car engines, transforming the battery's low voltage into thousands of volts, which in turn create an electrical spark that ignites the fuel.

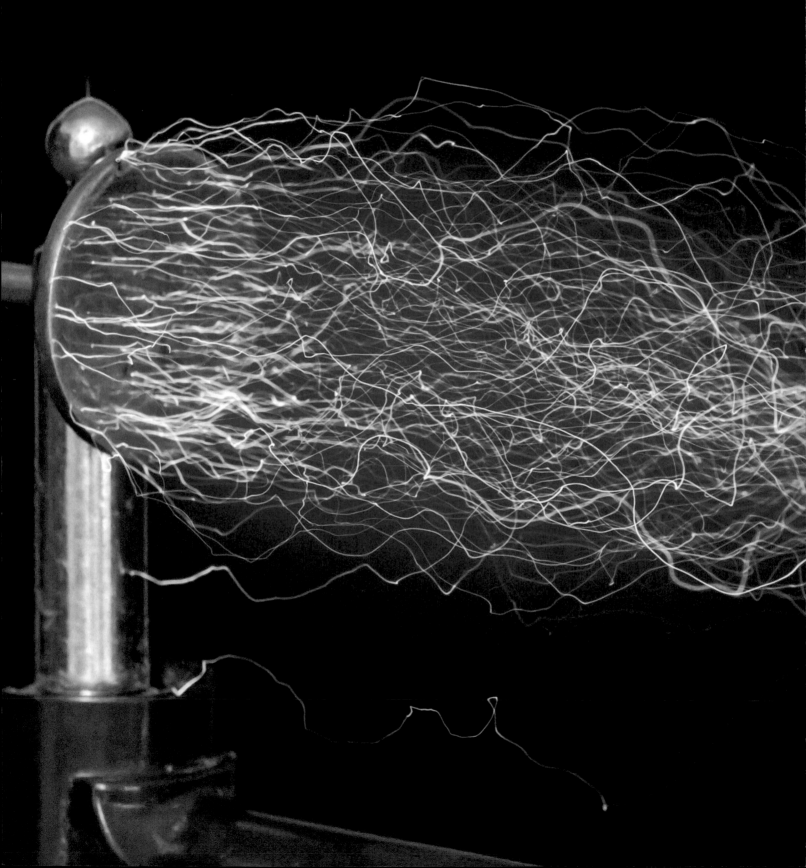

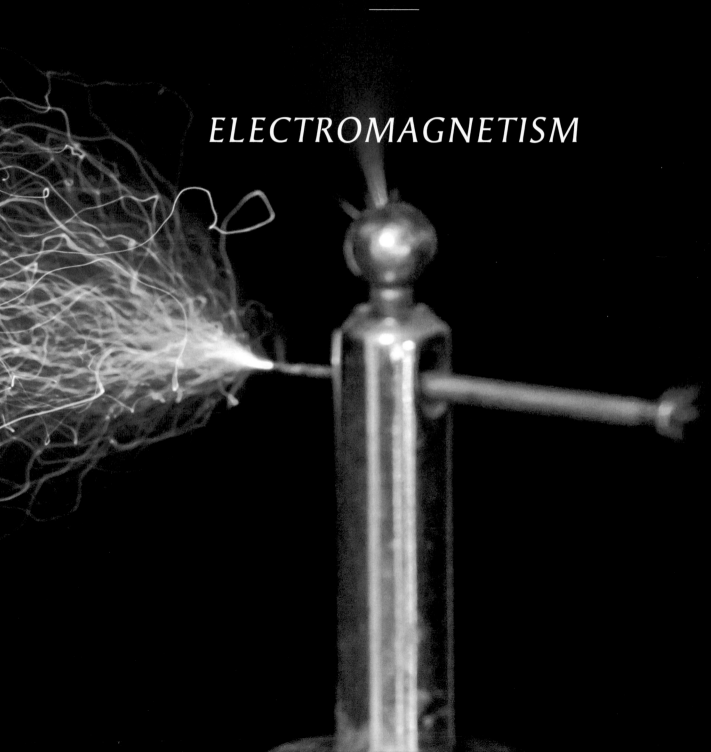

# CHAPTER

## 8

*ELECTROMAGNETISM*

## ELECTRICITY AND MAGNETISM

The title of this chapter—electromagnetism—is probably an unfamiliar word to you. It might sound like a special power a superhero possesses. But in fact, you are probably more familiar with it than you think—you have probably even felt it, like when you drag your feet on the carpet and then reach for a metal doorknob.

What happens?

You feel a slight electric shock! If you drag your feet enough, you might even see a little spark between your finger and the doorknob. Similarly, you may have rubbed a balloon on the carpet and then used it to raise a sibling's hair, or you can even "stick" the balloon to the wall.

The Greeks, many centuries before Christ, discovered a similar phenomenon. They noticed that they could rub an amber gemstone with fur to cause a spark. This was the beginning of human understanding regarding electricity, whose name comes from the Greek word for amber (*elektron*). Electricity is probably a word you are familiar with, but could you define it? A simple definition would be that **electricity** is a form of energy generated from negatively and positively (opposite) charged atoms reacting to each other. In our everyday lives, electricity is what provides power—power to cool your home in the summer, or to turn on your television.

But electric is only the first part of the word electromagnetism—the latter half of the word sounds like magnet, doesn't it? Perhaps you have magnets that stick to your refrigerator and hold things up. **Magnetism** is a phenomenon produced by the motion of an electric charge, resulting in an attraction (things coming together) or a repulsion (things pushing apart). All magnets have north and south poles—opposite poles are attracted to each other, while the same poles repel each other. We'll discuss this more later.

What people did not know until relatively recently in human history is that these two phenomena (electrical and magnetic charges) are related, and how they are related is a key part of our modern world. Together, they form the content of this chapter—**electromagnetism**—which is the science of charge and the forces and fields associated with it. Electricity and magnetism are the two areas that make up this scientific study. Let's now dive into each in more detail.

## STATIC ELECTRICITY

Electricity can either be stored as an *unmoving potential* or it can be moved in metal wires. The former is called **static electricity**, which is an electric charge that is not moving (static meaning stationary) but rather held in some object—like your finger after you drag your feet on the carpet. *Moving* electricity, however, is called an **electrical current** and it is made of *moving* charges that are used to do some useful work, like provide power to that television we mentioned a moment ago. Think of currents in

*Most of us are so used to power lines we fail to notice them, but they are vitally important to our way of life. They carry electrical currents to our homes and businesses, allowing us to power our appliances, televisions, and computers.*

the ocean that move like rivers—we spoke about them in our *Foundations* unit on oceans. Similarly, electrical currents move like rivers of electricity!

No matter the type of electricity we are talking about, we need a charge. When we say *charge,* just think of it as a burst of energy, like how you *charge* across the field when you are playing Red Rover. Scientists differentiate between two types of electric charges: one they call a positive charge, and the second one—you guessed it—a negative charge. The charges come from particles that have a built-in charge, which we will discuss when we talk about atoms. An object becomes *charged* when there is an unequal number of one type of charge particle compared to a second. So, if a particle is equally charged, but then it picks up a negatively charged particle, it will have a negative charge (it becomes *overloaded* with negative charges). This is what happens when you drag your feet on the carpet. The more of these particles you "pick up" (the longer and harder you rub your feet on that carpet), the stronger the static charge that is stored in your body to eventually be discharged when you reach out and touch that doorknob.

Objects that have the *same* charge will push each other apart, so two negatively charged objects will repel each other. Similarly, two positively charged objects will repel each other. On the other hand, two objects with *opposite* charges will attract each other (that's why we say "opposites attract").

*Static electricity is what makes your hair stand up when you rub a balloon against your head!*

# STATIC ELECTRICITY

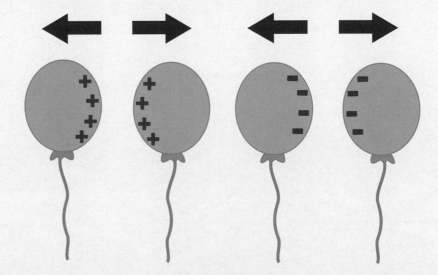

**Opposite charges attract**

**Like charges repel**

Charles-Augustin de Coulomb was an officer in the French military who was also a brilliant scientist. He lived in the eighteenth century. Coulomb noticed that the force of attraction or repulsion between the charges is proportional to the product of the charges (so the stronger the charge, the stronger the attraction or repulsion), and the force gets weaker the farther apart the charges are from each other. This should make sense. If two things are farther apart, they will struggle to make the same connection. This relation is called **Coulomb's Law** and the unit of measurement for charges is called the Coulomb. One interesting thing is that the force between the two charges is present even though there is no actual physical contact between them. You just have to get them close. We can actually measure the force and put a mathematical model to it, but there is nothing that touches the charges—the force is just there. If you rubbed that balloon on the carpet, you don't have to actually touch it to your brother's head to lift his hair, right? You just need to get it close.

With these basics behind us, let's talk more specifically about electrical currents, because so much of our modern lives depends upon them.

## ELECTRICITY IN MOTION

Once an electrical potential is present—once it becomes charged—it will try to return to a neutral state in which there is no net charge. Think of the particle as not liking the hyper or frenzied feeling it has, and it wants to discharge that feeling and return to normal (though of course particles aren't alive and can't think or feel like this). To do this, the charged particles will move. Think back to your "charged" finger—when you get it close enough to the doorknob, the electricity will move toward it. This movement of the charged particles is called an electric current. Materials that are good at conducting (moving) these particles are called **conductors** and those that prevent the movement of the charged particles are called **insulators**. Most metals are good conductors, while plastics are good insulators. If you notice, all the electrical cords in your house are well insulated to keep the charge moving as the engineers designed it to move. If you ever cut one of these cords (don't do that!), you would notice that

in the middle there is a metal core, which is where the electric current actually moves. For the current to exist, there must be something "pushing" it; this push is called the **electric potential**.

One of the great inventions of the early modern age was a machine that allowed scientists to create an electric potential. This could be attached to wire to create currents that do useful things. An Italian scientist named Luigi Galvani discovered, while dissecting frogs of all things, that if the metal of the dissecting tray and that of the scalpel he was using were different, an electric current would occur that made the

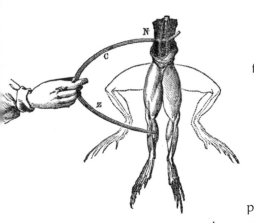

frog twitch. Alessandro Volta, another Italian scientist, used this information to create the primitive batteries called voltaic cells. The unit for electric potential, the **Volt** (V), is named after Volta. If you look at batteries around your house, you will notice they have their electric potential labeled (1.5 V, 3 V, 9 V, etc.); the larger the number, the more potential energy is stored in the battery to do something useful, like powering your remote-control car.

With the ability to control the voltage, scientists were then able to attach wires to both sides of the voltaic cell to create a closed loop for the charged particles to move through. This closed loop of current is called an **electric circuit**. These circuits are found everywhere in the modern world: in cell phones, computers, cars, refrigerators, televisions, toasters, stoves, and more!

## MAGNETISM

At least as far back as the Greeks, we have witness of "magical" stones that attracted ferrous metals to them naturally. These were eventually called **lodestones**, stemming from a Middle English word that meant "leading stone," because they were used in early navigation as the first magnetic compasses. The compasses were equipped with a magnetized needle that always pointed North (in the northern hemisphere), responding to Earth's magnetic pole. Lodestones are capable of transferring this "metal attracting property" to any metal that is in long contact with them, especially those made of a mineral called magnetite. We say that the lodestone can magnetize a second metal—in other words, it passes on that magnetic property.

### Batteries and Benjamin Franklin

Prior to Volta's invention of the voltaic cell, scientists stored electric voltage in glass jars with a liquid—water or alcohol—or a piece of metal. These were called Leyden jars. The jars had a metal piece coming out of the top that would be connected to a wire, which was itself connected to a machine that produced static electricity. As the static electricity was produced, the charge would move into the jar to be stored. Benjamin Franklin did research in which he connected several of these in series. He may be the first one to have used the word "battery" in reference to stored voltage. This is still the word we use for storage of voltaic potential.

*The steel of these paper-clips is magnetically attracted to this magnetite rock.*

These magnets have some interesting properties. Not only do they attract metals to themselves, but if you take two of the magnets and place them end on end, they will either attract or repel each other. The magnets seem to have two different ends to act—we call these **poles**. One pole is called north, and the other south. Same poles repel each other and opposite poles attract, similar to how charges behave. Another interesting thing is that if you cut a magnet in half, you do not cut the north pole from the south pole. The new smaller magnets would both have a north pole on the side of the original magnet's north pole and a south pole on the side of the original magnet's south pole.

Just like electricity, magnets are used in everyday life, and not just to hold things up on your refrigerator. They are also used in toys, jewelry, compasses, hospital and industrial machinery, and in many other ways.

## ELECTROMAGNETISM

Now that we have an understanding of electricity and magnetism, let's put them together and return to the concept of electromagnetism.

André-Marie Ampère, a French engineer, discovered that an electric current on a wire turned that wire into a little magnet. The magnetic effect is stronger with a stronger potential on the circuit. He also noticed that if he coiled the wire about a conducting metal, the effect became stronger the more loops he made with his wire. Then an English scientist, Michael Faraday, discovered that if you have loops of wire around a magnet and you move the magnet in and out of the loops of wire, an electric current is created whose potential increases with the number of wires and how fast the magnet moves in the coils of wire.

Most of the electrical production in the world uses these discoveries to convert mechanical energy, whether from a fossil fuel or wind turbine, into

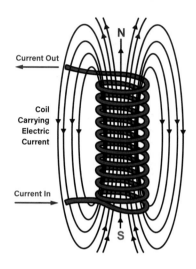

**MAGNETIC FIELD**

Current Out

Coil Carrying Electric Current

Current In

N

S

*An illustration of Michael Faraday's experiment using loops of wire around a magnet. When he moved the magnet in and out of the loop, an electric current was created.*

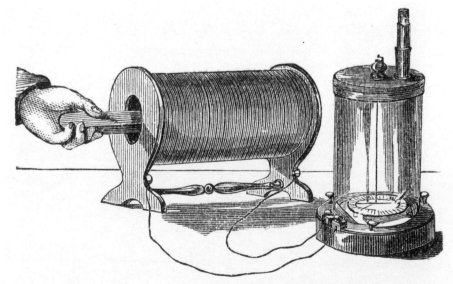

FIG. 358.—*Faraday's first Experiment.*

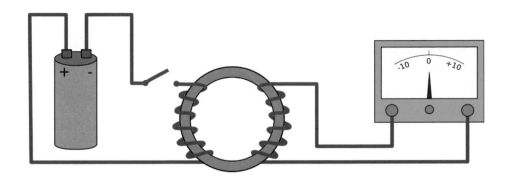

electrical kinetic energy that is then moved by wires into every house and business to power our modern lives. So, if you have lamplight shining on this book right now to help you read it, thank the phenomenon of electromagnetism and the scientists who discovered it!

Much of what we have discussed in this chapter, indeed in this entire book, is based on our understanding of atoms and their behavior. It is to the atomic model that we will turn in our next chapter.

## FOUNDATIONS REVIEW

✓ Electromagnetism is the science of charge and the forces and fields associated with it. Electricity and magnetism are the two areas that make up this scientific study. Electricity is a form of energy generated from negatively and positively (opposite) charged atoms reacting to each other, while magnetism is a phenomenon produced by the motion of an electric charge, resulting in an attraction (things coming together) or a repulsion (things pushing apart).

✓ Electricity can either be stored as an *unmoving potential* (static electricity) or it can be moved in metal wires through electrical currents. Both kinds require a charge. Some charges are positive, others negative. An object becomes "charged" when there is an unequal number of one type of charge particle compared to a second. So if a particle is equally charged but then it picks up a negatively charged particle, it will have a negative charge (it becomes "overloaded" with negative charges). Objects that have the *same* charge will push each other apart, but objects having *opposite* charges will attract each other.

✓ Magnets attract metals to themselves, but if you take two magnets and place them end on end, they will either attract or repel each other. This is because magnets have poles—one north and one south. Like electrical charges, same poles repel and opposite poles attract. Together with electricity, magnets form the science behind electromagnetism, which has many uses in modern life.

# Benjamin Franklin's Kite Experiment

One of the most well-known stories of Benjamin Franklin tells of when he flew a kite in a thunderstorm to prove that lightning is a release of static electricity. Franklin had hypothesized that lightning was an electrical discharge and that placing metal rods in strategic places could direct the discharge of lightning to protect buildings. He wrote a letter to Peter Collinson of the Royal Society of London stating his hypotheses. When Collinson read the letter to the society, Franklin's ideas were mocked and ridiculed. The kite experiment was Franklin's way to prove his hypotheses.

The actual experiment was a little more complex than the simplified version we usually hear about. Franklin built his own kite using silk (instead of paper) for it to withstand the wind of the thunderstorm. He attached a small metal rod to the top of the kite, and a rope to the kite, then attached a piece of silk to the string, so he was not touching the same string through which the electricity would be conducted (to protect himself from injury).

Finally, he attached a key at the bottom of the string and connected a wire from the key to a Leyden jar where the charge would be stored for later use in experiments.

When a thunderstorm finally came, he stood in a barn to keep both himself and the silk of the kite dry. As the thundercloud approached, the pieces of string on the rope stood on end. As the string got wet, the conduction of electricity started and the static charge was stored in the Leyden jar. He was not actually hit by lightning and successfully proved that a charge separation occurred in the area of the thunderstorm.

We now understand that lightning occurs when the difference in charge is powerful enough to cause the "spark." The lightning creates enough energy to produce the light and heat, and the heat then causes the air to expand; this causes the clap of thunder that always follows lightning. This story of Benjamin Franklin's kite experiment is a great example of forming a hypothesis and then testing it, showing the scientific method in action.

As we move toward a study of the Atomic Theory, our focus turns to the exciting world of chemistry! This 3D atom model depicts the interaction of the various particles that make up an atom.

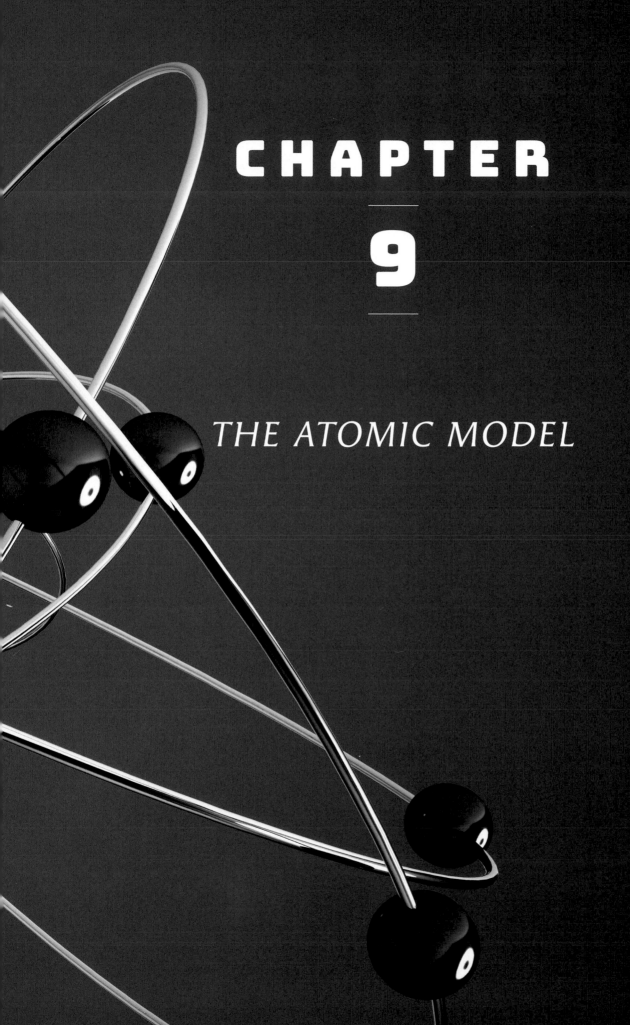

# CHAPTER
## 9

*THE ATOMIC MODEL*

## TURNING TO CHEMISTRY

Earlier in this book we spoke about motion, about things moving. The idea of motion in the pre-modern era included any and all changes that may occur in matter. For example, water evaporating was a "motion" of water as it changed from a liquid to a gas (water vapor). The change of pure iron to rusted iron was also considered a motion in that there was a change in the iron.

Up to this point, our discussion of motion has been confined to the changes in *position* of an object and the forces involved to move it from here to there, including the way energy moves. This is the definition of motion in the modern period, in our world. But the last four chapters of our book will introduce us to the science of chemistry, which studies the changes in matter as seen in the two examples above, and in many others.

## HISTORY OF ATOMIC THEORY

As we discussed in chapter 2, matter is classified into pure substances (elements and compounds) and mixtures. The study of chemistry helps us understand the motion of these substances, both those that are pure and those that are compounds. For example, iron, which is used in many tools, does not usually exist in its elemental or pure form. For it to be useful in making a screwdriver, it must be found in compound forms. These compound forms need to be changed to extract the pure iron from them. Ancient man learned early that heating the ores mined from the earth produced pure metals. Similarly, man knew by experience that grape juice was very sweet, but, if left to ferment, the sweetness would disappear and be replaced with the alcoholic taste, producing wine.

*When iron ore is heated up, it can give us important tools, like screwdrivers, that help us build things.*

Both of these are examples of chemistry at work. The first example of iron ore involves a compound being changed into an element, while the second one has a change of one compound (sugar) into a different one (alcohol). The first one involves heat, while the second one involves yeast cells eating the sugar to make alcohol.

# TIMELINE OF ATOMIC MODELS

| BILLIARD BALL MODEL | PLUM PUDDING MODEL | PLANETARY MODEL | BOHR MODEL | ELECTRON CLOUD MODEL |
|---|---|---|---|---|

| John Dalton 1808 | J.J. Thomson 1897 | Ernest Rutherford 1911 | Niels Bohr 1913 | Erwin Schrödinger 1926 |
|---|---|---|---|---|

The understanding of these changes is based on the atomic model of nature. The **atomic model** was first proposed by Democritus and Leucippus in ancient Greece as an attempt to understand the changes of matter seen in nature as one substance changes to another or as living things grow. They proposed that all matter was made of tiny, unbreakable components called *atomos*. The total amount of these *atomos* was considered finite, which meant that there was a finite (limited) quantity of matter in the cosmos. As they understood it, as an acorn grew into an oak tree it would "collect" more *atomos* to grow, and as it decayed many, many years later, those *atomos* would go back into the cosmos to be absorbed by other growing oaks.

This idea of the atom was forgotten until the early modern period when scientists adopted it in what they called *corpuscles*—this was what they used to refer to the basic smallest particles of matter and light. At this point, the theory was a philosophical idea and not an empirical, or observed idea, meaning they could not prove it, they were only speculating and tossing around ideas. These ideas would change as chemistry matured as a discipline and more precise studies were carried out.

## THE THREE LAWS OF ATOMIC THEORY

Three empirical laws were important to the development of atomic theory and each of them required very careful experiments. The first was the **law of conservation of mass** proposed by Antoine Lavoisier. This law states that matter can neither be created nor destroyed. Lavoisier performed many experiments and showed that if everything was carefully measured, all chemical changes involved no change in total mass. The amount of mass present at the beginning of the experiment was the same as the mass present after the chemical change took place. An example will help us understand this better.

Let's say you were to measure out a quarter cup of baking soda and you add it to four cups of vinegar in a beaker. The mixture will react and start to bubble and fizz producing small gas bubbles. If you did not collect the gas produced, it would seem as if there was a loss of mass. However, if you collected all of the gas produced in this reaction you would find that the mass of the vinegar and baking soda before the reaction is the same as the mass of the gases and the contents in the beaker after the reaction.

# LAW OF CONSERVATION OF MASS

*The law of conservation of mass states that matter can neither be created nor destroyed. This is seen in the diagram: even if the ice melts into water, the total mass is the same.*

**ICE CUBES**
**(100g)**

**LIQUID WATER**
**(100g)**

The second law of atomic theory is called the **law of definite proportions**, which states that all chemical compounds are made of whole number mass ratios of the elements in that compound, and that they are fixed and constant. This means that if we measure the mass of hydrogen and the mass of oxygen of any sample of water, we would find that there is one part hydrogen to eight parts oxygen (so water is ⅛ oxygen and ⅑ hydrogen). This is called the **mass ratio**. This is different from the chemical formula for water which describes the number of molecules it has, which of course is $H_2O$—two hydrogen molecules and one oxygen. What the mass ratio of water is in essence telling us is that, even though there are half as many oxygen atoms as hydrogen atoms in a molecule of water (1 versus 2), the oxygen has more mass, or takes up more space (eight times more). So all water, and only water, has this mass ratio of its elemental components of 1 hydrogen to 8 oxygen, and because of the law of definite proportions we know it will always be that way. This implies that all elemental particles of hydrogen and all elemental particles of oxygen have a fixed mass.

# LAW OF DEFINITE PROPORTION

**10 grams**
7.5 g C & 2.5 g H

**20 grams**
15 g C & 5 g H

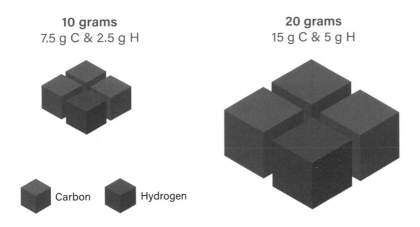

Carbon        Hydrogen

*If you have 10 grams of methane, 7.5 grams will be carbon and 2.5 grams will be hydrogen. If you have 20 grams of methane, 15 grams will be carbon and 5 grams will be hydrogen. Notice that the ratio of carbon to hydrogen remains the same. There is always 3 times more carbon than hydrogen by mass.*

The third law is called the **law of multiple proportions**, which states that if you have two different compounds made of the same elements, then they would have different whole number mass ratios. For example, hydrogen peroxide, like water, is made of only hydrogen and oxygen, but the mass ratio is one part hydrogen to sixteen parts oxygen. These two elements coming together to make hydrogen peroxide, then, would always have that ratio of 1:16. This is different from the mass ratio of water (a compound made of the same elements as hydrogen peroxide) which is one part hydrogen to eight parts oxygen. Here is another example. Carbon dioxide and carbon monoxide are both made of atoms of carbon and atoms of oxygen. In carbon monoxide there are 3 parts carbon to 4 parts oxygen, whereas carbon dioxide has 3 parts carbon to 8 parts oxygen. The compounds are different because they have different amounts of the same elements.

# LAW OF MULTIPLE PROPORTION

Octane
16:3

Propane
13:4

Methane
3:1

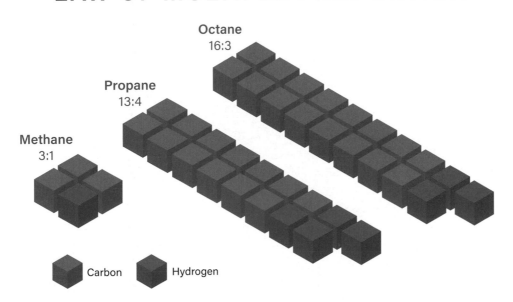

Carbon        Hydrogen

*Compare three different compounds that are made of only carbon and hydrogen. Methane gas contains 3 parts of carbon and one part of hydrogen by mass. Propane gas contains 13 parts of carbon to 4 parts of hydrogen by mass. Octane (gasoline) contains 16 parts of carbon to 3 parts hydrogen by mass. They are made of the same elements, but they have different mass ratios.*

## DALTON'S ATOMIC THEORY

These three laws were eventually combined into the first atomic theory by John Dalton, an English chemist (d. 1844). Dalton proposed the following principles for his theory:

1. *All matter is made of atoms which are very small and indivisible.* This means that all matter in the universe is made of these tiny bodies that cannot be seen even with a microscope. They are indestructible, meaning the atoms cannot be broken into anything smaller.

2. *Atoms of the same element are identical. Atoms of different elements differ in the mass they have.* This means that all atoms of the same element have the exact same mass, which explains the fixed ratios in compounds. If atoms have different mass, then they are different elements.

3. *All compounds are made of fixed ratios of atoms and different compounds with the same elements differ in their ratios.* This is a restatement of the two laws of proportions. This means that the ratio of hydrogen to oxygen in water is two atoms of hydrogen for each atom of oxygen—we write this as $H_2O$. For hydrogen peroxide, this ratio is two atoms of hydrogen to two atoms of oxygen, or $H_2O_2$.

4. *A chemical reaction or change involves the rearrangement of atoms.* In the example of the iron ore, this change means that the atoms of oxygen that were joined to the atoms of iron in the ore compounds are removed from the iron to make elemental oxygen and leave elemental iron behind. This is an example of a compound being broken down into its two elements. Rearrangements can also take different elements and combine them into a compound. When hydrogen gas and oxygen gas react to make water, the elemental form of hydrogen and that of oxygen rearrange into the atomic

## John Dalton: Father of the Atomic Theory

We tend to think of scientists as having a specialized education for a very long time, but in the early stages of the physical sciences, this was not the case. John Dalton is a good example of this. He was a Quaker who was not allowed to attend colleges in England because he was a Dissenter (English Dissenters were Protestant Christians who had separated from the Church of England through founding or attending their own churches). At the age of fifteen, he joined his brother in the running of a Quaker school. He got scientific instruction informally from John Gough, an English philosopher. By age twenty-seven, Dalton was appointed to teach mathematics and natural philosophy at the Manchester Academy, which he did for seven years, until he resigned due to the college's financial troubles. He then became a private tutor in mathematics and natural philosophy. He is well known for his work in meteorology, color blindness, and his contributions to chemistry. He is considered by many to be the "father of the atomic theory."

ratio found in the compound water. Since all matter is made of atoms, and atoms have fixed mass, then the rearrangement agrees with the law of conservation of mass.

At your age, this is a lot to take in. Don't worry if it doesn't all make perfect sense now. Just remember that this model was a great first step for chemistry, as it opened doors to systematic experimentation. With these laws established, scientists could carry out experiments and expect to see patterned results. Interestingly, though, with that experimentation, scientists discovered that the first two tenets of Dalton's theory were incorrect!

## DEVELOPMENT OF THE NUCLEAR ATOMIC MODEL

About six decades after Dalton, scientists invented and perfected an instrument known as the cathode-ray tube. They made a tube and removed the air from it to create a vacuum. They would partially fill the tube with a specific gas, say hydrogen, and then run an electric current through it. This allowed them to

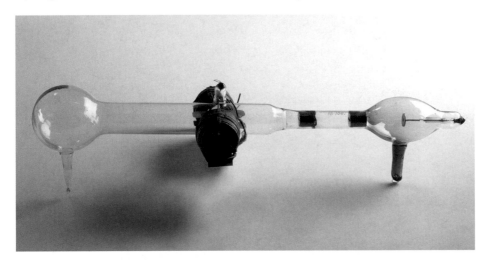

*The cathode-ray tube was used to discover the electron.*

observe a ray traveling from one side to the other (normally these rays would be invisible). This ray was shown to be deflected by a magnetic field, which implied that it is a moving charge (remember that electric currents create magnetic fields). Based on the deflection, scientists determined that it was a negative current. The explanation was given by J. J. Thomson, a British scientist, who proposed a negatively charged particle that he called the **electron**. He suggested that electrons were negatively charged particles found in atoms and that these electrons were inserted in a positively charged center, making it look sort of like a plum pudding. The elemental atoms are neutral, so the total negative charge and positive charge are exactly the same.

JJ THOMSON'S ATOMIC MODEL AND PLUM PUDDING

At about the same time, the French scientist Henri Becquerel discovered that salts of uranium would emit radiation without any external force. **Radiation** is energy put out by a source (in this case the element uranium) in the form of waves or particles; it is part of the electromagnetic spectrum we discussed earlier. Ernest Rutherford, a New Zealand physicist, discovered that the radiation came in three forms, which he called (1) alpha particles, (2) beta particles, and (3) gamma rays.

**Alpha particles** are heavy and have a positive charge equal in magnitude to two electrons. When collected, these alpha particles turn into helium as they collect electrons and become neutral.

**Beta particles** are negatively charged, with the charge equal to that of one electron, and they are much smaller than alpha particles.

Finally, **gamma rays** are electromagnetic waves of very high frequency. This suggested that not only do atoms have parts, but that they could break down or be divided into smaller elements, going against Dalton's first tenet. Rutherford also used alpha particles to discover that most of the atom is made of empty space. He found that almost the entire mass of the atom was found in a very dense core he called the **nucleus**. The electrons were all found outside the nucleus in what is called an **electron cloud**.

## EXPERIMENTS WITH RADIATION

The last piece of understanding of the nuclear atom came from continued experiments with radiation. In talking about radiation, we are once again using the technical definition of radiation instead of how we may typically think about it. In this instance, radiation, as explained before, consists of the alpha particles, beta particles, and gamma rays.

*The diagram below shows how some particles can penetrate certain surfaces but not others.*

## RADIATION *(ray)* PERMEABILITY

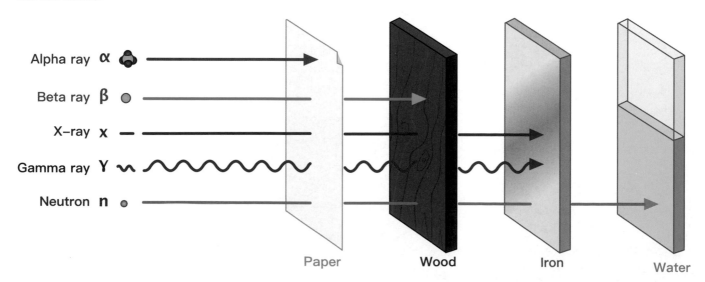

Alpha ray α

Beta ray β

X–ray x

Gamma ray Y

Neutron n

Paper   Wood   Iron   Water

*Pierre and Marie Curie are pictured here in their laboratory, c. 1900.*

Pierre and Marie Curie were the first married couple to win a scientific Nobel Prize, which they did in 1903. Marie was born in Poland but eventually moved to France to study at a university, where she met Pierre. Together, they discovered unknown elements that came from the radiation products of uranium salt.

Frederick Soddy, meanwhile, an English radiochemist who worked with Ernest Rutherford, whom we met a moment ago, found atoms of different mass that had the same chemical and physical properties, meaning they were atoms of the same element that had different masses. He called these **isotopes**. As a result of all of these experiments, we now know the following:

1. Atoms can break down or be divided;
2. Most of the mass of an atom is in the nucleus;
3. The electrons are outside the nucleus; and
4. Atoms of the same element can have different masses. We call these isotopes.

ATOMIC STRUCTURE

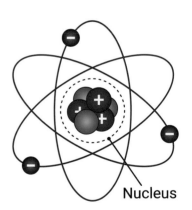

The current nuclear atomic model states that atoms are made of smaller particles called subatomic particles. There are three such subatomic particles:

• **Electrons:** These have a *negative* charge, and their mass is so much smaller than the other two subatomic particles (protons and neutrons) as to be considered negligible.

• **Protons**: These have a *positive* charge equal in magnitude to the charge of an electron, meaning that one proton will perfectly neutralize the charge of one electron. The protons have a mass about 10,000 times greater than the electrons. The mass of one proton is 1 AMU (Atomic Mass Unit). All protons are found in the nucleus of the atom. The number of protons also determines the type of atom, that is, the element of that atom. Hydrogen atoms have 1 proton, carbon atoms have 6 protons, oxygen atoms have 8 protons, and so on. So elements differ in the number of protons in their atoms.

• **Neutrons:** These particles have *no charge*, meaning they are neutral. They are a tiny bit bigger than a proton, but for our purposes we can say they have a mass of 1 AMU. They are found in the nucleus of the atom along with the protons. The total mass of the atom is the sum of the masses of all the protons and neutrons in the atom. Isotopes of the same element have the same number of protons but differ in the number of neutrons, resulting in different masses.

If that is a lot to remember, the most important thing is to know that electrons have a negative charge, protons have a positive charge (pro = positive), and neutrons have no charge (neutral = neither positive nor negative).

The current model we have of the atom helps characterize different types of atoms using something unchanging for that element: the number of protons. This explains the fact that atoms of the same element can have different masses as a result of their different numbers of neutrons. One difficulty of the model is that it concentrates all the positive protons together in the nucleus. As you might remember, equal charges repel each other, so how can this be? Scientists understand there is another force involved, which they call the **strong force**. This force counteracts, or works against, the electric repulsion of the positive charges of the protons. It requires neutrons in the nucleus of the atom to help "glue" the nucleus together. The electrons are in the "cloud" outside the nucleus, and it is these electrons that control the way atoms join together to make compounds. We will learn more about this in the next chapter.

*Remember:*

*Protons have a positive charge, electrons a negative charge, and neutrons a neutral change (no charge).*

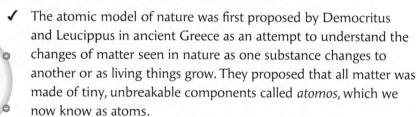
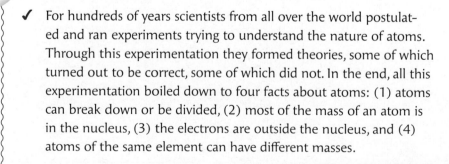

# FOUNDATIONS REVIEW

✓ The atomic model of nature was first proposed by Democritus and Leucippus in ancient Greece as an attempt to understand the changes of matter seen in nature as one substance changes to another or as living things grow. They proposed that all matter was made of tiny, unbreakable components called *atomos*, which we now know as atoms.

✓ For hundreds of years scientists from all over the world postulated and ran experiments trying to understand the nature of atoms. Through this experimentation they formed theories, some of which turned out to be correct, some of which did not. In the end, all this experimentation boiled down to four facts about atoms: (1) atoms can break down or be divided, (2) most of the mass of an atom is in the nucleus, (3) the electrons are outside the nucleus, and (4) atoms of the same element can have different masses.

# The Life-Giving and Destructive Qualities of Water

In this chapter we discussed the development of the atomic model of matter. This model has proven to be very robust in our understanding of the composition of material things and how this composition changes to make new compounds while conserving matter. This understanding of matter gives us a glimpse into what Aristotle called "the material cause," an understanding that was unimaginable before. This added structure allows us to understand the behavior of substances based on their material causes.

For example, water is made of one atom of oxygen bonded to two atoms of hydrogen. The shape of the molecule is understood by scientists based on the properties of the two atoms and how they come together. This, in turn, helps scientists understand why water is liquid at room temperature and why it freezes in colder temperatures, as well as how water dissolves some substances but not others.

Were we to describe water in less scientific ways, we may talk about how it feels or how it tastes. We also notice that it is essential for life, but that it can be destructive as well, as we see in canyons formed by water erosion or in the effects of great floods.

Water is also used by God in the sacrament of Baptism. As you know, a sacrament is an efficacious sign instituted by Christ to give grace. All sacraments have matter that is essential to them. For the sacrament of the Eucharist, wheat bread and grape wine are the physical materials needed, while water is the essential matter of Baptism. In this sacrament, the old self dies with Christ and a new self is "born" through God's grace. Notice that the sacrament uses both the destructive and life-giving properties of water. The old self dies in the waters, but we rise to new life in Christ.

The Periodic Table is the most recognizable symbol of the study of chemistry. It has changed and grown bigger over the last century and a half, now with 118 known elements.

81

$13^{3+}_3$   $11^{3+}_4$

$2s^2 2p^2$

$4, 2, -4$

77 2 d   77   170

2.0

$10^{-3}$

$_5 B$

Boron

$_6 C$

Carbon

[Ne] 3s

$4, -$

117

[Ne]

3

143.2 12

(3)

1.5

7.57

$_{13} Al$

Aluminium

IIIB

[Ar] 3d$^{10}$ 4s$^2$ 4p

# CHAPTER

## 10

## *ATOMIC BONDING AND THE PERIODIC TABLE*

**Proton**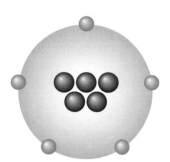

**Electron**

**Neutral Atom**

**CHARGED STATE**

↓

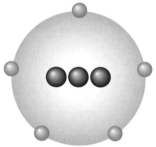

**Anion**
*(more electrons than protons)*

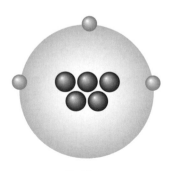

**Cation**
*(more protons than electrons)*

## CHARGED-UP ATOMS!

In the previous chapter, we looked at the modern atomic theory and how we understand the structure of atoms. We will start this chapter by looking at the different known elements. Remember that the original definition for an element is that it is a pure substance that cannot be broken down by chemical reactions. But the atomic definition of **elements** is that they are pure substances made of the same type of atom. Each element is determined by the number of protons in the atom's nucleus.

We usually think of atoms as having a neutral charge, which means that the number of electrons is equal to the number of protons. Think if you had a friend on the other side of a seesaw who weighed exactly the same as you—the seesaw would be in a neutral position (flat), neither up, nor down. It's kind of like that with atoms. This means that an atom of fluorine which has nine protons in its nucleus needs nine electrons in the electron cloud to equal the number of protons and make the atom "electrically neutral." Similarly, an atom of sodium which has eleven protons would need eleven electrons to perfectly balance the charge to produce a neutral sodium atom. I think you probably get the picture.

However, atoms are not always neutral, because many times in nature they are found in a *charged state*. These charged atoms are called **ions**. This state happens when the number of protons and electrons in the atom are *not* equal. This can only happen when an atom either loses or gains one or more electrons. Remember that the element of the atom is determined by the number of protons—*this means that the number of protons cannot change*. If you need help remembering that, perhaps you can think of protons like your fingers and electrons like your teeth. You can lose teeth but hopefully not your fingers!

If an atom loses electrons, then it has more protons than electrons and it is said to have a net positive charge—it is called a **cation** (pronounced cat-eye-on). An atom that gains electrons has more electrons than protons and has a net negative charge—it is called an **anion** (pronounced an-eye-on). Both of these are ions.

It may help to see it written out a little differently:

Cation = loses electrons → therefore has more protons than electrons → net positive charge

Anion = gains electrons → therefore has fewer protons than electrons → net negative charge

For example, sodium atoms tend to lose one electron and so are said to have a charge of positive 1 (making them cations). Fluorine atoms, meanwhile, gain one electron and so are said to have a charge of negative 1 (making them anions).

## THE HISTORY OF THE PERIODIC TABLE

As you probably know by now, scientists like to arrange things into groups to help keep track of them. You learned this in the *Foundations* unit on animals when we learned that biologists will group animals by like characteristics, giving us the subcategories of insects, birds, fish, reptiles, and mammals. In a similar way, chemists need to have an ordered system to keep track of elements and all their properties. This leads us to a discussion of the periodic table.

The first scientist to compile a list of the elements was John Dalton, whom we met in the last chapter. Dalton listed all twenty elements known at the time of his writing and arranged them by relative mass, having given hydrogen a mass of 1 AMU. In the years that followed, scientists continued to find new elements in nature to add to the list. By the end of the 1860s, chemists had found sixty-three different elements!

A Russian chemist, Dmitri Mendeleev, noticed that there was a pattern to the elements as he listed them by increasing mass; eventually, he would get to an element that behaved chemically very similarly to a smaller element. It also had similar physical properties, like two brothers or two sisters looking and acting alike. For example, both sodium and potassium react with oxygen in a 2:1 atomic ratio. They are also soft, shiny metals that are not naturally found in their elemental forms. Understanding what these similarities are is not the important thing here, but rather that Mendeleev noticed the pattern. He added rows to the list of elements and placed similar elements in the same column, stacking them on top of each other, just like you might stack same-colored

JOHN DALTON'S ELEMENTS

*The white boxes on this Periodic Table represent the missing elements Mendeleev knew existed in the world, they just had not been discovered yet.*

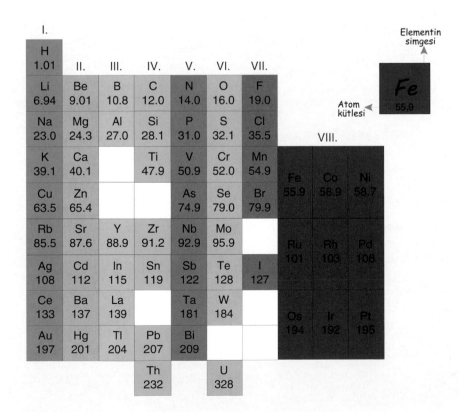

blocks on top of one another. As he did this, he left gaps where an element seemed to be missing because there was no known element that fit the pattern. In other words, nothing was there to fit the pattern . . . *yet*, but Mendeleev knew there should be. Within a decade, two such elements had been found with the masses he had predicted. We call this arrangement a periodic table because the repeated properties are said to occur in cycles, or periods. The elements in columns are in the same family and behave similarly, and the rows are the different periods.

The modern **periodic table** is considerably larger than Mendeleev's, as there are now 118 known elements! The elements are currently arranged not by weight/mass as they once were but by the number of protons, which is called the **atomic number**. This is a good feature to organize them by because, remember, the number of protons cannot change. Of these 118 elements, only 94 occur naturally on Earth, while the rest are man-made. The periodic table has a square for each element—at the center of each square there is a symbol or abbreviation for the element (H = hydrogen, O = oxygen, etc.) with the element's full name underneath it. The atomic number (the number of protons it has) is in the top of the square.

# PERIODIC TABLE OF ELEMENTS

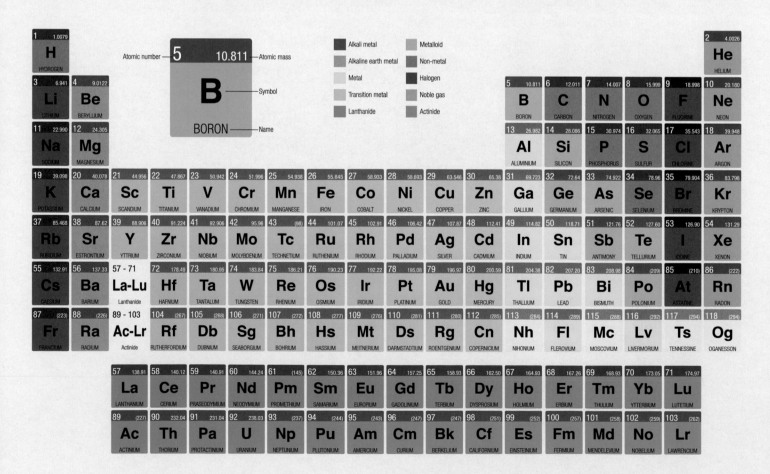

## Organic Chemistry

Carbon is a very important element for life. It is found in 99% of all known compounds and all the biological molecules are formed on a carbon backbone. Sugars, DNA, RNA, proteins, and fats are all carbon-based compounds. What makes carbon unique is that it only forms compounds by sharing electrons—this allows it to make large molecules of bonded carbons. Graphite and diamonds are examples of molecules made entirely of carbon. Carbon's chemistry is so important that there is a whole field of chemistry on the study of carbon called organic chemistry.

*Chemistry Fun Fact:*
*Hydrogen, helium, oxygen, carbon, and neon are the five most common elements in the Milky Way Galaxy.*

Most known elements are classified as metals, as shown in the periodic table in this chapter. Notice that the nonmetals are found on the right side of the periodic table. Generally speaking, metals are good conductors of electricity, they are spongy and malleable (flexible/changeable), and, when found in compounds with nonmetals, they tend to lose electrons and make cations. Nonmetals tend to be nonconductive and, when they react with metals, they tend to gain electrons and make anions.

There are a few elements that have properties of both metals and non-metals (these are shown in teal on the table). These are called **metalloids**. The most important of these in our modern world is silicon. Metalloids are semi-conductors, which means that they only conduct electricity under some conditions. This property is what allows silicon to be the key ingredient in computer chips. You can probably guess why silicon is so important considering how much our modern world relies on computers.

## ATOMIC BONDING

When atoms come together to form a compound, we say that they bond with each other. A **chemical bond** is the force of attraction between two distinct atoms to make a new entity. There are two main types of chemical bonds that chemists speak of: ionic bonds and covalent bonds. The bonded atoms are expressed in a chemical formula that gives the whole number ratios of the elements in the compound. It is written with the atomic symbols from the periodic table with numbers written as subscripts. Let's look at each of these more closely, along with some examples.

As implied in the name, an **ionic bond** involves ions of opposite charges that attract each other to make a new compound. The compounds always have a net *neutral* charge. So if there are two positive charges, there will be an equivalent two negative charges to give a neutral charge. A compound made from ionic bonds is called an **ionic compound** and is commonly called a salt or mineral. The sodium cation and fluorine anion we discussed at the beginning of the chapter

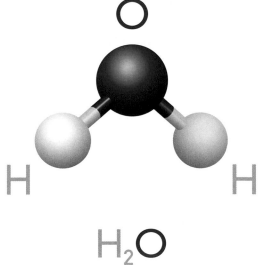

$H_2O$

CHEMICAL FORMULA

form an ionic compound in which the positive sodium ion and the negative fluoride ion attract each other to make the salt sodium fluoride, with the formula NaF. The fluorine is renamed fluoride in its anion form and it is found in your municipal water and many toothpastes.

If an atom loses two electrons, then it would have a charge of positive 2. Magnesium and the elements below it in the periodic table behave in this way. It would take two fluoride ions to get a total of 2 negative charges to balance the magnesium's charge, so the formula for the magnesium fluoride compound would be $MgF_2$.

Most known compounds do not form ionic bonds. They form a type of bond known as a **covalent bond**. This is a bond in which, instead of gaining or losing electrons, the atoms "share" the electrons between them to form a bond that keeps them together. Over 99% of all known compounds are these covalently bonded compounds, and most of these are made just of carbon, hydrogen, nitrogen, oxygen, phosphorus, and sulfur. Scientists define atoms that are covalently bonded to each other as **molecules**. Sometimes we get atoms of the same element bonded to each other to make what would be called **molecular elements**. The oxygen gas you breathe is a molecule made of two oxygen atoms covalently bonded to each other, so it has the chemical formula $O_2$. This is a pretty important molecular element if you like breathing! Another example would be the graphite found in pencils—it is made from very large chains of carbon atoms covalently bonded to each other.

Molecules are made of whole atoms which is why we find whole atom ratios in chemical compounds. For example, a molecule of water is made from two hydrogen atoms sharing electrons with one oxygen atom and, as we noted previously, has a chemical formula of $H_2O$. A molecule of glucose is made from six atoms of carbon, twelve atoms of hydrogen, and six atoms of oxygen, sharing electrons in a very particular way and forming a very specific

*Remember:*

*Carbon is the primary building block of all living organisms.*

# COVALENT BONDS

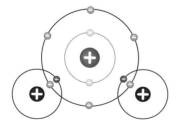

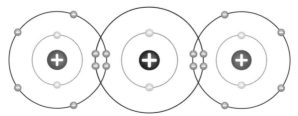

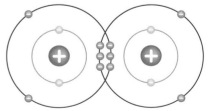

H — O — H ⟶ $H_2O$
**SINGLE BOND**
**IN**
**WATER MOLECULE**

O = C = O ⟶ $CO_2$
**DOUBLE BOND**
**IN**
**CARBON DIOXIDE MOLECULE**

N ≡ N ⟶ $N_2$
**TRIPLE BOND**
**IN**
**NITROGEN MOLECULE**

structure. The chemical formula for glucose is $C_6H_{12}O_6$. The study of carbon containing compounds like glucose is called organic chemistry, a more specific branch of chemistry.

Carbon is a very unique element in that it very rarely forms ions; it is almost exclusively found in covalently bonded compounds. Carbon, as in the graphite in your pencils, is very good at forming covalent bonds with other carbon atoms, allowing for long chains to form. Thus, it is easy to see why carbon is the main building block of living organisms.

With this understanding behind us, we can now move on to discuss chemical reactions in our next chapter.

## FOUNDATIONS REVIEW

✓ If an atom has the same number of protons and electrons, it has no charge (it is neutral). But atoms are not always neutral, because many times in nature they are found in a charged state. These charged atoms are called ions. This state can only happen when an atom either loses or gains one or more electron(s), because the number of protons cannot change. If an atom loses electrons, then it has more protons than electrons and it is said to have a net positive charge—this is called a cation. An atom that gains electrons has more electrons than protons and has a net negative charge—this is called an anion.

✓ The modern periodic table compiles 118 known elements and arranges them by the number of protons they have, which is called the atomic number. Of these 118 elements, only 94 occur naturally on Earth, while the rest are man-made. The periodic table has a square for each element—at its center, there is a symbol or abbreviation for the element (H = hydrogen, O = oxygen, etc.) with the element's full name underneath it. The atomic number is in the top of the square.

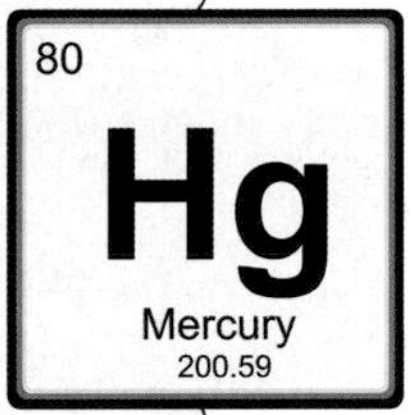

80
**Hg**
Mercury
200.59

✓ When atoms come together to form a compound, we say that they bond with each other. A chemical bond is the force of attraction between two distinct atoms to make a new entity. There are two main types of chemical bonds that chemists speak of: ionic bonds and covalent bonds. Ionic bonds involve ions of opposite charges that attract each other due to their opposite charges to make a new compound, while in covalent bonds there is no loss or gain of electrons; rather, the atoms "share" the electrons between them to form a bond that keeps them together. Over 99% of all known compounds are covalently bonded compounds.

# Entropy: Physical and Spiritual

In chapter 6 we introduced entropy from the perspective of mechanics. In mechanics, entropy is looked at as unavoidable waste. The atomic view of material creation provides a better understanding of entropy and why it is unavoidable.

Imagine a small box with sixteen Legos. Since each of them is a separate unit, there are a lot of ways to arrange them around the box. As you assemble them into larger pieces, you organize their structure such that there are fewer ways to arrange them in the box, since some Legos are now attached to each other. If you attach all sixteen of them together, then you have maximized how organized they are and, thus, decreased entropy (disorder) in the box. In other words, there is now structure to it. This analogy can be used to gain a glimpse into the atomic view of entropy.

Entropy is a measure of molecular disorder: the higher the amount of disorder, the higher the entropy. According to Newton, atoms are seen as always in motion. The smaller the molecules (groups of atoms) are and the faster they are moving, the higher the quantity of disorder, sort of like how a scene of ants scurrying around on an ant pile looks chaotic. But as molecules get bigger (like the Legos attaching to one another), the amount of entropy is decreased. Living things are highly ordered with very large molecules working together in a cell; and, for multicellular organisms, the cells are organized together. Remember that a law of thermodynamics states that the entropy of a closed system (e.g., the universe) always increases. How, then, do living organisms stay alive if entropy must increase?

We can start with a second analogy. Have you noticed that your play area tends to get messier the longer you play without picking up all the toys? This is an increase of entropy over time. What

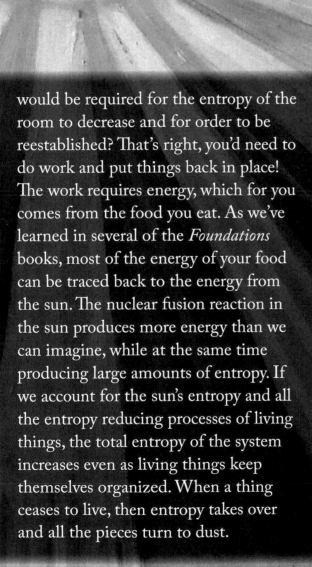

would be required for the entropy of the room to decrease and for order to be reestablished? That's right, you'd need to do work and put things back in place! The work requires energy, which for you comes from the food you eat. As we've learned in several of the *Foundations* books, most of the energy of your food can be traced back to the energy from the sun. The nuclear fusion reaction in the sun produces more energy than we can imagine, while at the same time producing large amounts of entropy. If we account for the sun's entropy and all the entropy reducing processes of living things, the total entropy of the system increases even as living things keep themselves organized. When a thing ceases to live, then entropy takes over and all the pieces turn to dust.

Spiritually speaking, grace is the energy that helps us order our lives to God's will. As long as we stay in a state of grace and keep cooperating with actual graces, our lives will be ordered to God and the fruits of the Holy Spirit will take over. If all humans were working out their salvation with fear and trembling, an ordered society would exist. Sadly, original sin introduced spiritual entropy (disorder and chaos) into the world. Our actual sin increases this entropy, and the combined sinfulness of humans leads to a disordered society that needs God's grace to regain order. This is why the world we live in is full of such pain and sadness.

But we must never despair, because where sin abounds, grace abounds all the more (see Rom 5:20)!

This thermite explosion is one of the more dramatic chemical reactions you will find!

# CHAPTER
## 11

## CHEMICAL REACTIONS

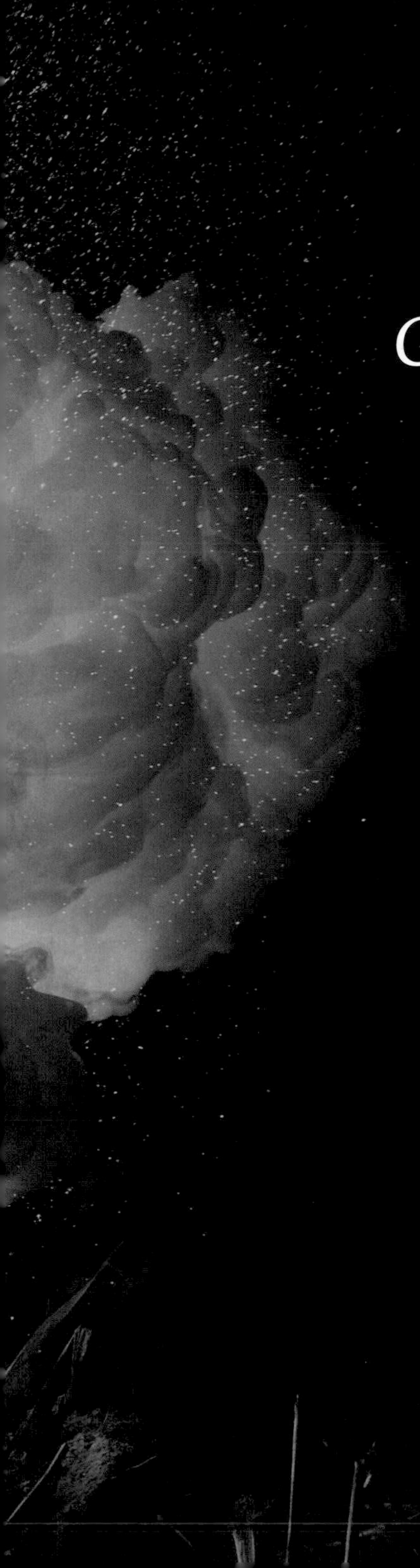

## REACTANTS INTO PRODUCTS

John Dalton's atomic theory was a great starting point for the scientific under-standing of the atomic model of matter. Over the years, chemists found that real atoms were not as he imagined, but his ideas of chemical compounds (the laws of proportions) and chemical reactions are still used in the understanding of chemistry.

If you recall, Dalton defined a chemical reaction as the rearrangement of atoms, similar to when you rearrange the bricks of a Lego creation into a different form. We say that a chemical reaction starts with a number of chemicals, called the **reactants**, or reagents. In the process of the chemical reaction, the initial chemicals become new chemicals, which are called the **products**. If you have gotten far enough in your math studies, you know that the answer to a math multiplication problem is called the product—it is the answer to whatever the numbers multiplied together equal. In a similar way, the product of a chemical reaction is what comes after the reaction takes place. Thus, we start with reactants and get products as the atoms rearrange. Chemists have become quite proficient at investigating and controlling chemical reactions.

The law of conservation of mass, which states that mass is neither created nor destroyed, can be extrapolated to chemical reactions in that the number of atoms that are rearranged in a chemical reaction will be the same at the end of the reaction as before the reaction. In other words, the number of atoms in the reactants will be equal to the number of atoms in the products. In the equation $2 + 2 = 4$, you have the same amount on either side of the equal sign, right? It is similar to that. We can also say that we balance the mass involved in the chemical reaction by balancing the number of atoms involved in the reaction.

## CHEMICAL EQUATIONS

We already used math equations to help you understand chemical reactions better. Well, there is actually something called a chemical equation too. Scientists are very precise in the way they communicate how chemical reactions occur. For example, we can say that hydrogen, $H_2$, reacts with oxygen, $O_2$, to make water, $H_2O$. The reactants are $H_2$ and $O_2$. The product is $H_2O$. The statement is true, but it does not convey the whole picture of what is going on. It is not precise enough.

The first thing to note is that the molecule of hydrogen has *two* atoms of hydrogen, and the molecule of oxygen has *two* atoms of oxygen. But the product has *two* atoms of hydrogen and only *one* atom of oxygen. Do you notice the problem? The number of oxygen atoms in the reactants and products do not match. There are two oxygen atoms in the reactants, but only one oxygen atom in the product.

How do we solve the problem?

Well, we can say that one molecule of oxygen containing two atoms of oxygen ($O_2$) is going to make *two* molecules of water, but that requires *four*

*Chemistry Fun Fact:*

*Pure iron reacts with oxygen to make rust. This is why playground equipment made of iron begins to rust after years spent out in the open air (out in the oxygen), along with the rain that makes it wet.*

atoms of hydrogen (2 atoms from the first molecule and 2 atoms from the second molecule). This means that we need two molecules of $H_2$ to get a total of four atoms of hydrogen (2x2=4). Are you following us so far? If not, take a look at the image on this page. It shows this with the reactants on the left and the products on the right. Each blue sphere is an atom of hydrogen, and each red sphere is an atom of oxygen. If you count below, you will notice that there are four blue spheres (hydrogen atoms) on the left and four on the right. Similarly, there are two red spheres (oxygen atoms) on the left and two on the right.

# CHEMICAL REACTION

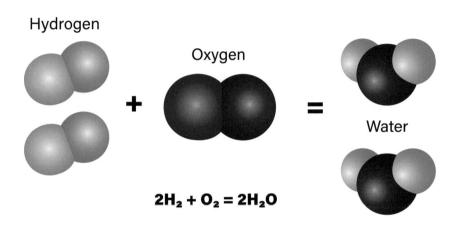

Hydrogen

Oxygen

Water

$$2H_2 + O_2 = 2H_2O$$

*This diagram shows the chemical reaction that gives us water. Notice how the reactants (left side) and the products (right) are equal.*

This gives most of the information scientists wish to convey. The only thing that is missing is the state of matter of the species as the reaction takes place. In this case, all three species are gases (abbreviated g), and the reaction takes place in the gaseous state, which will be discussed in the next chapter. Drawing pictures of the reactants and products is not efficient, so scientists use chemical formulas instead to simplify the information. All this is used to write a chemical equation that expresses the chemical reaction. For our water example, the equation is:

$$2H_2(g) + O_2(g) = 2H_2O(g)$$

This reads: Two molecules of hydrogen gas react with one molecule of oxygen gas to produce two molecules of water gas.

## REDUCTION-OXIDATION REACTIONS

Scientists have many ways of describing chemical equations depending on the type of changes that occur. One of these changes is called an **oxidation reaction**. If you notice, *oxidation* has the same root as *oxygen*. In fact, oxygen means "oxide maker." Initially, oxidation reactions were those that had oxygen as a

## CHEMICAL COMPOUNDS OF RUST

**Oxygen**
O₂

**+**

**Iron**
Fe

**+**

**Water**
H₂O

↓

**Rust**
Fe₂O₃H₂O

reactant. For example, pure iron reacts with oxygen to make rust. This is why playground equipment made of iron, like the chains of swings, begins to rust after years spent out in the open air (out in the oxygen), along with the rain that makes it wet. It is probably also why playground equipment now tends to be made not of metals like iron but of high-density plastics.

We say that rust is the oxidized form of iron (iron + oxygen = oxidized iron, or rust). It would take two atoms of iron to react with one molecule of oxygen, along with water, to make two iron oxides (in this case, the iron and iron oxide are in the solid state, abbreviated s). The formula would look like this:

$$2Fe(s) + O_2(g) = 2FeO(s)$$

The process that occurs in plant cells using glucose and oxygen to make water is another example of an oxidation reaction. One molecule of glucose, $C_6H_{12}O_6$, reacts with six molecules of oxygen to make six molecules of carbon dioxide ($CO_2$) and six molecules of water (here water is in the liquid state, represented by an l).

$$C_6H_{12}O_6(s) + 6O_2(g) = 6CO_2(g) + 6H_2O(l)$$

These formulas are probably making you go cross-eyed! Don't worry about remembering them. For now, just understand the basic principle that when oxygen is added to something, or when it becomes mixed up with something else, it changes the physical makeup of what is present (like when iron begins to rust). Just think of it like baking a cake—the ingredients of sugar, flour, chocolate, eggs, etc., all have their own separate properties, but when they are mixed together and baked, they turn into a cake. Their physical composition has been changed into something quite delicious!

In the previous examples, we talked about oxygen being *added* to something else. But the opposite can also occur—oxygen can be *removed*. When we start with a compound with oxygen and then remove the oxygen, it is called a **reduction reaction** (reduction meaning to take away). For example, iron ore, when heated, produces oxygen gas and iron metal. This specific type of reaction is also called a decomposition reaction.

$$2FeO(s) = 2Fe(s) + O_2(g)$$

Notice that the reaction is the exact opposite of the first example we looked at—the formula for oxidized iron (rust).

$$2Fe(s) + O_2(g) = 2FeO(s)$$

# REDUCTION REACTION

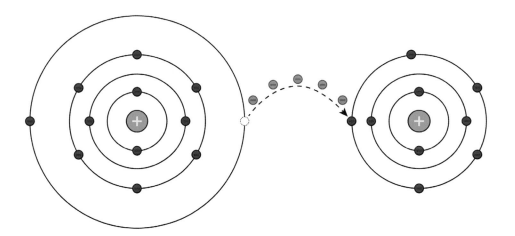

*This diagram demonstrates the transfer of an electron during a reduction reaction.*

As stated earlier, scientists have become very good at understanding and controlling chemical processes.

Again, the goal is not to memorize these equations at your age. The goal is to understand the reactions in a general way. Scientists say that all oxidation and reduction reactions involve both the oxidation and reduction process. In both cases, there is one element that is oxidized and one that is reduced. The oxidized element *loses* electrons and the reduced element *gains* electrons. In the case of the iron reaction with oxygen to make the iron oxide, chemists can trace the electrons as having moved from the iron, which lost them, to the oxygen, which gained them. The iron is oxidized, and the oxygen is reduced. This may sound very complicated, and the details are beyond the scope of this text, but these types of reactions are essential to life as we understand it and to the modern world in which we live.

## It All Starts with the Sun

Reduction-oxidation reactions are very common in living organisms. Plants, for example, perform photosynthesis to make sugars, which they use for energy and structure (it's their way of "eating" to get energy, but, unlike us, they make their own "food"). Photosynthesis is a series of reduction reactions performed on carbon dioxide to make sugars using sunlight. The process stores chemical energy by making a large molecule of sugar and producing oxygen gas as a side product. When animals (including us!) eat, they extract these molecules to mine energy to stay alive. They do this by a series of oxidation reactions that extract the energy from the larger molecules and produce carbon dioxide. The oxidation process uses molecules of oxygen to produce the carbon dioxide needed by plants to make the sugars most of the rest of the food web needs. In this way, most of the energy in the Earth's ecosystem begins with the Sun.

## COMBUSTION REACTIONS

An important subset of the reduction-oxidation reaction is the combustion reaction. **Combustion reactions** are those in which a "fuel" is oxidized with oxygen gas. Generally speaking, the fuels are carbon-based compounds like natural gas or propane. The combustion reaction produces carbon dioxide and water from the oxides or carbon and hydrogen found in the fuels.

Why are these so important to our everyday living? Well, all combustion reactions produce heat, which is useful in cooking, staying warm in winter, and in the conversion to electricity in power plants. They are particularly useful because one of the reactants (oxygen) is essentially free, as it is found in the air, so we need not store it along with the fuel.

## ENERGY IN REACTIONS

As was mentioned in the example of the combustion reactions, chemical reactions involve changes in energy. The rearrangement of atoms in the reaction causes changes in energy. A reactions that gives off energy (most often in the form of heat) is called an **exothermic reaction**. The heat *exits* the reaction as if it was a product, like your printer spitting out a piece of paper. Some reactions only occur when energy is added to the atoms to make the process happen. We looked at an example of the reaction to take iron ore and get the iron metal from the ore by *heating* the ore. This type of reaction is called endothermic. An **endothermic reaction** is one that requires that energy be added to it for the reaction to occur.

How do we understand these energy changes? We go back to the idea of potential energy. In an exothermic reaction, the reactants have more potential energy than the products. That energy is stored in the covalent bonds between the atoms. When the process happens—because energy is not destroyed—the

*A factory worker stands before a giant furnace used to create steel, one of the most common materials in the modern world, used in the production of cars, planes, trains, tools, and buildings. Steel is known to be 1,000 times stronger than iron!*

**ENDOTHERMIC**

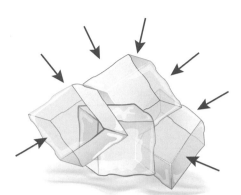

Absorb energy

**EXOTHERMIC**

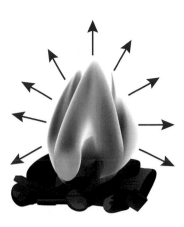

Release energy

energy exits the molecule in some form (usually as heat). In an endothermic process, the reactants have less potential energy than the products, so in order to make the reactants into the products we must add the energy difference from outside the atoms. In this way, we talk of chemical potential energy. Compounds with high chemical potential energy are useful as fuels, which is a key resource in our lives.

This discussion of energy is a good transition to our final chapter, which will discuss the fascinating world of the energy found in reactions.

## FOUNDATIONS REVIEW

✔ We say that a chemical reaction starts with a number of chemicals, which we call the reactants, or reagents. In the process of the chemical reaction, the initial chemicals become new chemicals, which are called the products. The number of atoms in the reactants will be equal to the number of atoms in the products.

✔ Oxygen is a common molecule that is involved with chemical reactions on Earth. It can either be added to another element or removed—these are oxidation reactions and reduction reactions. These types of reactions are essential to life as we understand it and to the modern world in which we live.

✔ An important subset of the reduction-oxidation reaction is a combustion reaction. Combustion reactions are those in which a "fuel" is oxidized with oxygen gas. Generally speaking, the fuels are carbon-based compounds like natural gas or propane. All combustion reactions produce heat, which is useful in cooking, staying warm in winter, and in conversion to electricity in power plants.

# Where Does Electricity Come From?

As human beings we have the ability to enhance our physical capabilities using tools. What we cannot do on our own, we can often do through the use of tools, like slamming a nail into a piece of wood using a hammer. The making of tools requires some form of energy; for handmade tools, the energy would be the food the toolmaker eats. One could not make a tool to complete a task without the energy to do it, which one gets from eating food.

For many years throughout our history, all we had was our own energy to get things done. But as time marched on, humans eventually started to use the energy from domesticated animals (a mule pulling a plow) and moving water (water mill) to increase our power over nature. With the technological advances of the twentieth and twenty-first centuries, humans can harness large quantities of energy stored in fuels and convert them into electricity.

As we saw in this chapter, large molecules tend to have high potential energy, which may be released through a chemical reaction. The main reaction used by man historically to release this stored energy is combustion. If you have ever been by a bonfire, you have felt this release of energy in the form of heat. For most of human history, fire has been the source of heating and cooking used by humans. In more recent times, we have used electricity to cool, heat, cook, and power many of our great machines.

Where does most of the electricity come from? Electricity is a form of kinetic energy, so it is not the source of the energy itself. The electricity we get has come mostly from converting the chemical potential energy of our fuels using the discoveries in electromagnetism we discussed earlier. A fuel is combusted to get water steam very hot, and, at a very high pressure, this steam is used to spin a coil of wire which is surrounded by a magnetic field. As you might remember, a changing magnetic flux produces an electric current. The spinning wire is where the electric current is made. So the potential energy of the fuel is converted to heat energy in steam, which is converted to mechanical energy to spin a coil of wire in a magnetic field, which in turn produces the electric current that is connected to most of our houses.

Water can be found in three states of matter—liquid, solid, and gas—depending on its temperature. No matter what, though, it is still water!

# CHAPTER

## 12

*ENERGY IN REACTIONS*

## STATES OF MATTER

Here in this final chapter of our unit, we will look at the nature of matter in the different ways in which a substance may be found. You have no doubt seen ice melt into water. And this water, if heated, can be turned into steam. All three—ice, water, and steam—are the same substance—water—but differ in the state or phase in which they are found.

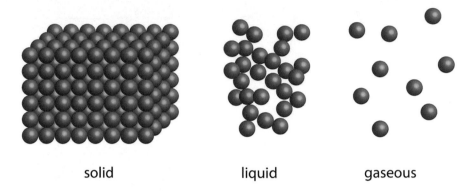

solid                liquid                gaseous

There are three primary states of matter: solid, liquid, and gas. The change from one phase to the other is not a *chemical* change like we discussed in our last chapter. Instead, it is a *physical* change. The substance, in this case water, remains the same, but is found in three different physical forms. We heat ice and in doing so turn it into water, and if we continue heating it, it will eventually boil into the gaseous phase of water, steam, which is also called water vapor.

Based on this process, can you guess which state of water has the highest kinetic energy? Remember kinetic energy is the energy of objects that are in motion and/or doing work. Let's look more closely at each of these three phases and perhaps by the end of this chapter you will be able to answer that question.

## GASEOUS STATE

The gas state is the *least* dense phase of matter. At the molecular level, gas molecules are very far apart from each other and constantly in motion. Think back to our example of the two suitcases when we discussed density—one with thirty T-shirts packed in it, one with three T-shirts. The suitcase that has thirty T-shirts is packed tighter, right? The less dense one has three T-shirts, so in essence, the T-shirts are farther apart and can jostle around inside when the suitcase is moved (there is more empty space in the suitcase). It is similar to the gas molecules that are less dense—they are farther apart and constantly in motion.

Imagine if it were possible to look into a balloon to observe what the air blown into it is doing. Inside we would see molecules of air constantly zooming past each other, colliding with each other, and colliding with the sides of the balloon, like a bunch of Ping-Pong balls bouncing around. Each collision of a molecule with the balloon would push the balloon with a force proportional to the size and speed of the molecule. All the collisions that are constantly happening *add up to a force* that acts over the surface area of the balloon, swelling it out. The ratio of force divided by area (the surface of the balloon) is called **pressure**.

## Force / Area = Pressure

All gases exert pressure on the container they are in, and this pressure is the same everywhere in the container. This air pressure is what gives the balloon its overall shape. You may also have experienced air pressure in your bike tire. The bike tire must contain a certain amount of air in order to maintain its shape (its firmness), allowing you to ride your bike. If there is a small hole in the tire, air leaks out, meaning there is less pressure being exerted on the tire. When this happens, it goes flat, and that is the end of your fun day of riding bikes with your friends.

You may not think of it like this, but the air in the atmosphere is constantly pushing down on your skin—we refer to this as atmospheric pressure. If you are at sea level, every square inch of your skin experiences 14.7 pounds of weight from the atmosphere. But as you go up in elevation and the column of air above you becomes smaller, you experience less air mass pushing down on you and the atmospheric pressure decreases. This is why the air is "thinner" in the mountains.

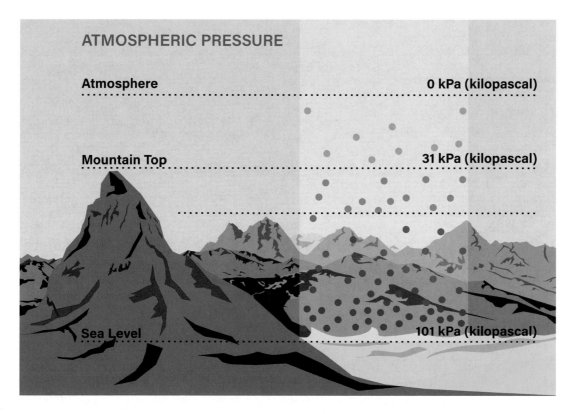

**ATMOSPHERIC PRESSURE**

**Atmosphere** .................................................. **0 kPa (kilopascal)**

**Mountain Top** .................................................. **31 kPa (kilopascal)**

**Sea Level** .................................................. **101 kPa (kilopascal)**

Returning to our balloon, if you move the air within it to a different container, you have not lost any molecules of air, but it is now in a different container with a different shape. If we change the shape of the container, the shape of the air is different, since all gases take the shape of the container in which they are found. If you have a container with a gas, it is relatively easy to push in on the gas and make the volume it occupies smaller. It is much harder to squeeze a solid rock than a balloon filled with air, right? If you do squeeze the balloon, this has the effect of increasing the pressure produced by the gas. In a sense, the balloon becomes smaller as you squeeze it, and so you will have more collisions within the smaller surface area of the smaller volume. This is called **compression**. Gases are the only state of matter that can be compressed a significant amount, so we say the gas phase is "compressible." Of course, if you compress the balloon too much, it will pop!

There are other ways to change the state of the gas. If we remove energy from a gas by cooling it, we would eventually force it to transition into the liquid state in a process called **condensation**. One place you can observe this is looking at a glass of cold water on a hot, humid day. As the water vapor in the air makes contact with the cold glass, the water vapor changes from its gas state to its liquid state, causing liquid water to form on the outside of the glass. The opposite is also true. If we take a liquid and add energy to it (add heat), then we turn it into a gas in a process called **vaporization**. The boiling of water into steam is an example of vaporization with which you are probably familiar.

*The liquid water that drips off your cold glass on a hot summer day is a result of the process of condensation, while the steam that rises from a heated teakettle filled with water is called water vapor (gas).*

If we look at the direction heat flows, we will notice that the gas phase has more energy than the liquid phase. Think about it—as we added energy to the liquid (when we heated it up), it became a gas. Conversely, when we removed energy from the gas (when it was cooled down), it became liquid. This shows us that gases have more energy than liquids. The temperature at which the change from liquid to vapor happens is called the **boiling temperature**. This temperature can be different for each substance. Thus, water has a different boiling temperature than that of natural gas, propane, alcohol, and other substances.

## LIQUID STATE

In the liquid state the molecules of the substance are now close enough to attract each other. This makes the space between molecules close enough that they act differently in liquid than in a gas (the T-shirts in the suitcase are now closer together). That being said, liquids do share some properties of gases. Like a gas, a liquid will take the shape of the container in which it is placed. That's obvious, right? If you pour water into a glass of water, the water will take the shape of the glass. Also, like gases, a liquid can flow over or across a surface. You can feel both the wind flowing over your skin on a windy day and the flow of water in a stream. These properties make the gas and liquid phases of matter *fluid* phases.

Unlike the gaseous phase, though, the liquid phase is very hard to compress. Imagine trying to squeeze a full water balloon into a smaller space. If you try hard enough, you will most likely end up wet, as the balloon pops due to the force of water pushing out. As there is little leftover room between liquid molecules, it takes a lot of energy to be able to force them to get closer to each other. For practical purposes, we say that the liquid phase is "incompressible." This means that if we apply a force to a container with a liquid, that pressure will be the same everywhere in the container. This can be used to transfer the force to another part of the fluid container. The incompressible property of fluids was used in hydraulic steering and hydraulic brakes in older cars and trucks.

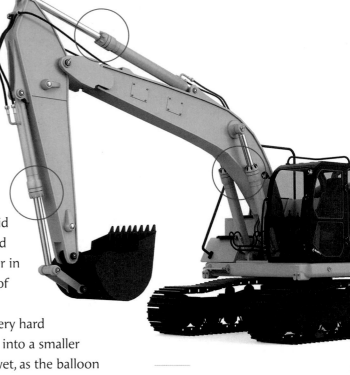

*Believe it or not, water is used by means of hydraulic pressure to move parts of this giant excavator.*

## The Uniqueness of Water

As a liquid gets cooled and its molecules slow down, most solids form a crystal structure with its particles being closer to each other than they were in the liquid phase. This means that the same quantity of a substance occupies *less* space in the solid phase than it does in its liquid phase (it takes up less room). As a substance transitions from the liquid to the solid phase, the bits of solid sink in its liquid, as they are denser than the liquid. One very important exception to this process is water. Due to the structure of the water molecule, as water crystals form, the distance between the molecules increases and ice becomes *less dense* than liquid water. This means that solid water floats in its liquid, which is why ice floats in a drink. Out in nature, when ice forms on the top of a lake, the ice insulates the water and prevents the whole lake from freezing. This is a very important exception, as it allows life underwater to survive in the depths of winter and thrive in the coming warm seasons.

If a liquid is cooled enough, then the substance will eventually become solid in a process called **solidification**. In contrast, a solid that is heated enough becomes liquid in a process known as **liquefaction**. The freezing and melting of water are common examples of these processes. Which of these two phases has a higher energy state? The liquid state has higher energy than the solid because the latter has energy *added* in the form of heat to liquefy it. The temperature at which this happens is called the *melting point* or the *freezing point*. Of course, these two are the same temperature; the difference is that the melting point refers to the transition from solid to liquid and the freezing point is the transition from the liquid to solid. This temperature is particular to different substances, just like we saw with the boiling point. The state (solid or gas) the liquid transfers to depends on the direction in which the energy is moving (are you heating it up or cooling it down?).

## SOLID STATE

The solid phase of a substance is usually the densest phase of matter of a substance (water being a very significant exception). Usually the solid particles form in a regular arrangement known as a crystal. Solids keep their own shape, are even less compressible than liquids, and do not flow. All that should make sense if you just think about what a solid is. You cannot compress a brick and change its shape, and it cannot "flow" over you like a gas or liquid can.

There are two main types of solids depending on the type of compound they're made from. An ionic compound forms an ionic solid. The crystals of ionic solids are hard and have very high melting points. The other type of solid is made of molecular compounds (remember molecular compounds are those that are held together by covalent bonds). These are called molecular solids. Molecular solids can be very soft, almost waxy, like petroleum jelly, and are generally softer than ionic solids. They also have considerably lower melting points.

DRY ICE

We have looked at the transition of solids to liquids, which is the transition we most often observe. But it is possible to transition directly from a solid to a gas phase (in other words, to skip over the liquid phase). This change is called **sublimation**. The main example you might be familiar with is the transition of dry ice to a gas. Dry ice is solid carbon dioxide and it must remain very cold. It goes from the solid phase to the gas phase at -78.5 °C.

The hardest solid we know of is a form of carbon we call diamonds. Your mom probably loves these! Diamonds are neither ionic nor molecular solids. They are an example of what is called a network solid. Network solids are made of very large numbers of atoms covalently bonded to each other. Diamonds are large networks of covalently bonded carbon atoms. Perhaps, if you save up enough money you can buy your mom a diamond necklace for Mother's Day!

## THE MOLECULAR VIEW OF TEMPERATURE

At the beginning of the chapter, we asked which phase of water has the highest level of kinetic energy. Remember, in a previous chapter we mentioned that heat is a form of kinetic energy. In this chapter, we discovered that the gas state has more energy than the liquid, and liquid has more energy than the solid. We know this due to the heat changes that must occur to make the phase changes happen. We must add heat energy to melt ice into liquid water, and more heat is needed to boil the liquid in order to form the vapor. How, then, would one understand the relationship between the temperature of a substance and the kinetic energy it has?

Remember that the kinetic energy of an object increases as the object moves faster, just like you are exerting more energy if you are running around versus lying on a bed. In relation to heat content, we have stated that an object at a higher temperature has more energy than one at a lower temperature. Putting these two ideas together with the molecular understanding of matter, we can say that at higher temperatures the molecules are moving faster. In a gas the molecules move much, much faster than in a liquid, and thus the kinetic energy difference between the two states is very large. Liquid molecules move faster than solid molecules, but the difference is lower than that between gases and liquids. This means the energy difference between liquids and solids is not as large as that between liquids and gases.

*Remember:*
*Sublimation, though not as common as solidification and liquification, can still occur. This is when a solid is heated to such an extent that it skips over the liquid phase and goes straight to a gas. The most common example is dry ice.*

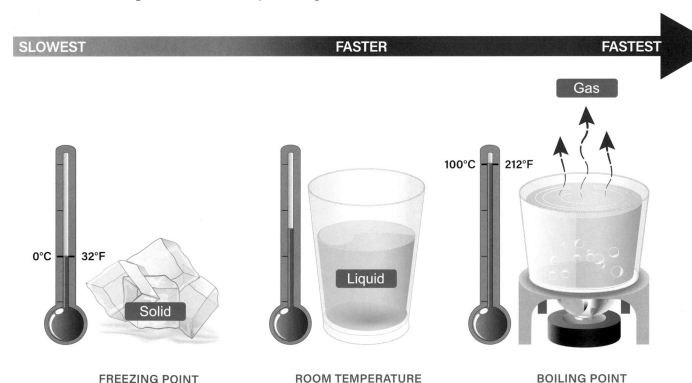

SLOWEST          FASTER                    FASTEST

Gas

100°C — 212°F

0°C — 32°F

Solid

Liquid

**FREEZING POINT**          **ROOM TEMPERATURE**          **BOILING POINT**

It is difficult to say that at a given temperature a given substance has a given quantity of kinetic energy, or velocity, as the numbers are affected by the shape, structure, and other factors of the substance. But we can say that as the temperature increases, the average velocity of the particles of that substance increases, and, as a result, the kinetic energy increases. The opposite is also true. With all of this is mind, hopefully you have concluded that it is the water in the gas phase which has the highest level of kinetic energy!

Well, it has been a long and fruitful journey through the world of physics and chemistry. To conclude our unit, flip ahead to read some final thoughts.

## FOUNDATIONS REVIEW

✓ There are three primary states of matter: solid, liquid, and gas. The change from one phase to the other is not a chemical change, but a physical one. In the case of water, it can be found in ice, liquid water, and steam or water vapor. We heat ice and in doing so turn it into water, and, if we continue heating it, it will eventually boil into steam.

✓ The gaseous state of matter is the least dense and the most compressible, while the solid state is the densest and the least compressible, with liquid being between them. Changes from one state of matter to the next occur as energy (heat) is either applied to or removed from the matter. So liquid water becomes ice when heat is removed (when it is made cool), and it becomes a gas (steam) when heat is applied to it.

✓ The gaseous state has the most kinetic energy, and the solid state has the least. We know this due to the heat changes that must occur to make the phase changes happen. We must add heat energy to melt ice into liquid water, and more heat is needed to boil the liquid in order to form the vapor.

# Conclusion

We have finished our survey of the physical sciences! Along our journey, we looked at the ways we understand the operation of the physical world. This understanding has given us a power over nature that was unimaginable to people hundreds of years ago.

And yet, this is not the end point of the understanding, but rather this study should elevate our eyes to the source of the whole of creation. The material world, even as we understand so much about it, does not explain itself. In other words, nowhere in nature does something create itself—it must always come from something else. So you come from your parents, and an oak tree from an acorn, and a flower from a seed. Even an infinite series of efficient causes leaves unexplained the source of the first existence. Taking this into account, St. Thomas Aquinas wrote that there must exist some being that exists on its own, that fully explains itself. This is God.

We also saw that in the nonliving world we get glimpses of purpose and intentionality, even when we are unable to understand what something's final cause may be. These glimpses, like ice floating on liquid water, point to a source of final purpose to all of creation. This, once again, is God. The contemplation of the natural world should be a starting point in contemplating the greatness and awesomeness of God, not an end to itself.

We hope that this book has taught you much about the physical sciences, but more than that, we hope that it is a launching point for you to study the natural world even more, and, in doing so, to draw you closer to the Creator who rests behind it all.

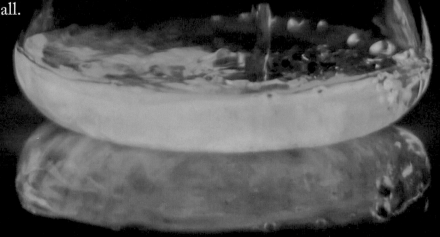

# AMAZING FACTS ABOUT CHEMISTRY & PHYSICS

- Even though a bowling ball and a balloon filled with air have approximately the same volume (size), they have vastly different densities, as the bowling ball is far, far heavier (it has more mass packed into that volume). This is a good comparison for understanding the difference between volume and mass, which gives an object its density.

- Though we can use our senses to take in the various forms of matter in the world, for the most part, only some senses can discern matter at one time. Some things you can smell but you cannot see, or we can feel air when it is blowing, but our eyes do not see it. Still, matter is discernable to our senses, even if not to all of them.

- Everything you have studied in the *Foundations* series is made of matter, from the largest star to the smallest bacteria, from a blue whale to a grain of sand. This means studying chemistry and physics leads us to the study of *everything*.

- The highest of the sciences is theology because it is the study of Him Who is the highest subject of study: God. The second highest science is philosophy as it studies the created world. What we now call "science" was known as natural philosophy as it searches for truth about the *material* world, in other words, the stuff made of matter.

- Humans have used measurements in trade for many, many centuries. In England, people used human scale measurements: number of feet between distances, number of strides to measure larger measurements (the yard), number of hands to measure how tall a horse is (the unit of hands), the amount of land an ox and one man could plow in one day (the acre). But these can vary from person to person, so more uniform measurements were eventually established (inches, meters, etc.).

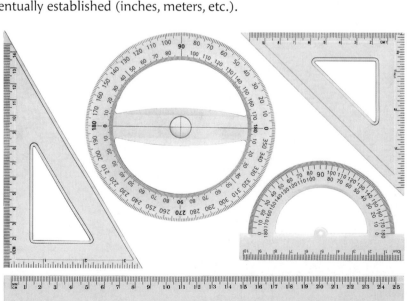

- It is important to note the difference between a hypothesis and a theory. The former is an educated guess based on observations and initial measurements; the latter is supported by repeated experimentation and evidence.

- The earliest study of motion began with attempts to understand the movement of the planets in the night sky.

- Most world-class sprinters cross the finish line near the highest instantaneous speed, at about 12 $^m\!/s$. To think of that in a more familiar way, that means they are running 12 meters in a second, which is almost 40 feet!

- When you feel movement—like when you get on a rollercoaster—what you feel is the change in motion, or the acceleration of the object. An object accelerates when its speed changes, when the direction of movement changes, or when they both change. When a car slows down or increases speed, you can actually *feel* the change. But you may also feel the acceleration of an object whose speed does not change, like when a car makes a turn without changing speed. A better example is on an airplane that cruises at constant speed and then makes a turn. You didn't really feel the movement of the plane until it started to turn, though its speed stayed the same.

- Sir Isaac Newton was an English mathematician, physicist, and astronomer living in the seventeenth and eighteenth centuries. He is remembered today as one of the most influential scientists in history, primarily because he proposed three laws that form a predictive model for the motion of objects—this means the laws help us predict what will happen when it comes to the motion of an object. Newton's proposals completely changed the way scientists look at the physical world.

*Physics Fun Fact:*
*The earliest study of motion began with attempts to understand the movement of the planets in the night sky.*

- The second law of motion states that the acceleration of an object of constant mass is proportional to the force applied on the object. This means that the larger the force applied to a given object (the harder the pitcher throws a ball or the batter hits it), the greater the acceleration of the object. You could also say, if you desire a greater acceleration of the object, a greater force is needed to cause the acceleration. This is why bat speed (how quickly the batter swings the bat) is a key factor in successful baseball players, those who hit well and hit many homeruns. The faster you swing the bat, the farther the ball will go.

- The third law of motion is known as the action-reaction law. It states that when a force is applied to an object, the object applies a *reaction force* equal in size to the force applied on it—but in the opposite direction. This means that as a chandelier pulls down on the stud in the ceiling, the ceiling is applying an equal force as it pulls up on the chandelier. Or, as you push on the wall, the wall "pushes back" at you with the same force.

- When you push a toy car on the floor, the force of friction acts on the car to slow it down until it eventually stops. The amount of friction between the floor and the car changes when we change the type of surface that the car rolls on. A car pushed with equal force will roll farther on a smooth, wooden or marble floor than it will on carpet. The understanding is that the carpet has more friction than a smooth floor. Generally speaking, a rougher surface produces more friction than a smoother surface.

- Friction is essential to our being able to walk or to drive a car. If you have ever found yourself on a very slick surface, you probably noticed that walking becomes quite difficult. We see the same thing when streets become very icy and it is hard for a vehicle to get going or to stop. As you walk, you push on the ground, and the ground pushes back on your foot. It is the reaction force of the ground that propels you forward and allows you to move. If there was no friction, then there would be no reaction force and you could not move forward. The slickness of the ice reduces the friction below your feet, making it harder to push off and walk (or to drive a car).

- Drag through air is called air resistance, or aerodynamic drag, while the force of drag through a liquid is called hydrodynamic drag (hydro = water). Air resistance slows down a Frisbee on a windy day, and walking through a pool is harder and slower than walking on land because of hydrodynamic drag.

- Drag is a function of the shape of the object and the density of the fluid through which it is passing. The force of drag increases as the fluid gets denser. Water is much denser than air, therefore it has much more drag than air. This is one reason why airplanes can fly faster than submarines can move. The force of drag also increases the faster the object moves. This means a car moving at 20 mph experiences significantly less drag than one moving at 70 mph.

- Carmakers do a lot of research to optimize the shape of their cars to decrease the amount of drag, as this helps the cars to have better aerodynamics, and, as a result, use less fuel.

- Earth obviously has a lot of gravity since it is so large, so it pulls us down to the planet's surface. Technically, as the planet pulls us down to it, we also pull on the planet with our own gravity; but since we are so very small in relation to the mass of Earth, there is no effect on Earth's motion. In fact, even if all humans alive today were in the exact same spot, we would still have no effect on Earth.

- Interestingly, if we were on an object with less mass than Earth (if we were on a smaller planet), then our weight would immediately be lighter than it is on Earth. For example, the Moon has one-sixth (⅙) the mass of Earth, so your weight on the Moon would accordingly be one-sixth (⅙) what it is on Earth, due to the smaller gravitational force experienced. Conversely, your weight would increase on a larger planet like Jupiter.

- When you throw a baseball from the outfield to home plate, you notice that it makes an arc. Mathematicians call this shape a parabola. It starts to go up and forward until it reaches a high point—or apex—and then starts falling until it hits the ground or is caught by someone on the ground. The ball does not go up indefinitely because it experiences the pull of gravity from Earth to come back down to the surface. You can change the shape of the parabola by changing the initial angle in which you throw the ball. The same thing is experienced by anything that moves through the air.

• If a shuttle is moving fast enough and gets high enough, it will enter the weightlessness of space, where there is much less gravity (although not none, as some people mistakenly believe). The shuttle will enter an orbit where it is, in a sense, constantly falling and constantly missing the Earth. Astronauts in orbit experience the state of weightlessness because they are constantly falling, not because the Earth has ceased to have an effect on them.

• The *higher* the object, the *higher* the amount of potential energy that can be converted to kinetic energy. A great example of this is water stored in a water tower that needs to be pumped to a town. The water is stored at an elevation because that height can be used to turn the water into moving water, to eventually provide water service to the townspeople. Another example is a hydroelectric dam which turns a column of water into an electric current.

• The heat capacity of a substance is how easily some mass of the substance changes temperature—how quickly it heats up or cools down. Materials that readily get hot or cold are called conductors. They are used when heat needs to be moved; for example, we use metal to make cooking pans because we want the heat from the stove to readily transfer to the food to cook it. Materials that do not easily change temperature are called insulators and are used when we need to keep heat in or out of somewhere. For example, if you need coffee to stay hot for a long time, you place it in an *insulated* container which prevents the transfer of heat. An insulator in this instance would also protect your hand from getting burnt while you hold the cup, whereas you need an oven mitt to hold a hot pan (a conductor).

• When you drop a stone into a pond, if we could slow down time and zoom in, we would see that the ripples resulting from the splash have peaks where the water is higher and valleys where it is lower, like a circular mountain range.

Both the sound you hear and the ripples you see moving away from the point of impact are ways in which the kinetic energy of the rock is *dissipated* in both the air and the water—meaning the energy is scattered out.

- Even things you may not think of as waves, such as earthquakes, are mechanical waves. The energy that causes the quake—the disturbance of tectonic plates grinding against one another—is dissipated through Earth's crust in what are called seismic waves. Mechanical waves move at different speeds and, generally speaking, move faster through denser things, such as solid objects.

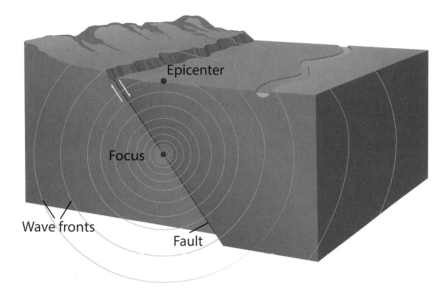

- The electromagnetic wave you are most familiar with is the one that allows you to read this text: light! Microwaves, radio waves, UV-rays, and X-rays are all examples of electromagnetic waves. Unlike mechanical waves, all electromagnetic waves move at the same speed when in a vacuum.

- Sound moves on average at about 340 $^m/s$ through the air—this means it's traveling just over 1,000 feet in a second! If you think that's fast, though, light travels in a vacuum at an astonishing 300,000,000 $^m/s$—that's almost a million times faster than sound! Since light is so much faster than sound, this explains why during a thunderstorm you see the lightning before you hear the accompanying thunder boom. Or if you are watching a golfer strike a golf ball from far away, you will see him swing first and hear the pop of the club against the ball a second or two later.

- If sound waves move at the same speed, why doesn't everything sound the same? Well, because what makes the different pitch of the sounds you hear are not differences in wave speeds but in the frequency of the waves. Your ear is capable of hearing sound waves between 20 Hz and 20,000 Hz. At lower frequencies you hear low-pitched sounds—think of the piano keys at the far end that make an ominous thunder-like *boom*. But as the frequency increases, the pitch of the sound increases—you will hear this in the higher notes on the opposite end of the piano, notes that sound almost like a teakettle whistling.

*Remember:*
*A seismic wave is a wave in the Earth's crust caused by an earthquake.*

- The loudness of the sound is related to how much energy is being moved by the wave. If you barely tap the table, you will hear a very faint sound. If you hit the table harder, then the sound is louder since you gave it more energy. When you increase the volume of your sound system, what you are doing is increasing the energy used by the system to make sound waves.

- For sound to echo, the sound waves must hit a hard surface so that they bounce back. Softer surfaces tend to absorb the sound so that there is no echo. If you were to yell in a large, empty gymnasium, you may hear the sound of your voice echoing back. But, if you are in a movie theater with soft carpet and hundreds of comfortable chairs, the noise gets absorbed and does not echo back.

- Resonance refers to when waves of the same frequency vibrate with each other, and their effects get amplified, or added to one another, and thus they sound louder. If you were to remove a guitar string and pluck it in the middle of an empty room, then it would vibrate in place and make only a faint sound. But once you place the string in a guitar's body and pluck it, the sound is a lot louder as the waves in the cavity of the guitar (the inside of the wooden part) add their effects and resonate with each other to give a louder sound.

- Light waves are electromagnetic waves with wavelengths between 400 nm and 700 nm (nm = nanometer, one billionth of a meter). This is the whole range of light that is visible to the human eye—anything less or anything more and we cannot see it unless we have some instrument to help us. Different colors have different wavelengths, but when the whole spectrum of light is seen together, we see white light.

- A white-colored object is one that is reflecting the entire spectrum of light, while a black-colored surface is absorbing the whole spectrum of light. This is why it is better to wear white in the summer when the sun is intense—it reflects more of the light so that your body does not absorb as much heat/energy.

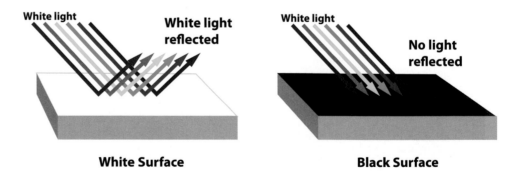

- The green grass at the park is reflecting green light to your eyes and absorbing the other light colors. You might think of it as the grass is taking in and "hiding" the other colors from you (absorbing them), which leaves only the green light for it to show.

• When the light travels from one medium to another and its speed changes, the path of the light bends in the direction of the denser medium and this is called refraction. You can see the effect of refraction if you look at a straw or spoon in a glass of water. From the side, it may look like the straw or spoon is broken or bent. Refraction is studied using glass or plastic prisms that can be used to bend light. If light is bent enough then it disperses into the different colors of the light spectrum, making a rainbow. This is because the amount of bend is slightly different for the different colors of the spectrum—different degrees of bend reveal different color shades of light.

• The word "electricity" comes from the Greek word for amber (*elektron*).

• An object becomes *charged* when there is an unequal number of one type of charge particle compared to a second. So, if a particle is equally charged, but then it picks up a negatively charged particle, it will have a negative charge (it becomes *overloaded* with negative charges). This is what happens when you drag your feet on the carpet. The more of these particles you "pick up" (the longer and harder you rub your feet on that carpet), the stronger the static charge that stored in your body to eventually be discharged when you reach out and touch that doorknob.

• Once an electrical potential is present—once it becomes charged—it will try to return to a neutral state in which there is no net charge. Think of the particle as not liking the hyper or frenzied feeling it has, and it wants to discharge that feeling and return to normal (though of course particles aren't alive and can't think or feel like this). To do this, the charged particles will move. Think back to your "charged" finger—when you get it close enough to the doorknob, the electricity will move toward it. This movement of the charged particles is called an electric current.

- Materials that are good at conducting (moving) these particles are called conductors and those that prevent the movement of the charged particles are called insulators. Most metals are good conductors, while plastics are good insulators.

- If you cut a magnet in half, you do not cut the north pole from the south pole. The new smaller magnets would both have a north pole on the side of the original magnet's north pole and a south pole on the side of the original magnet's south pole.

- Most of the electrical production in the world uses these discoveries to convert mechanical energy, whether from a fossil fuel or wind turbine, into electrical kinetic energy that is then moved by wires into every house and business to power our modern lives. So, if you have lamplight shining on this book right now to help you read it, thank the phenomenon of electromagnetism and the scientists who discovered it!

- Pierre and Marie Curie were the first married couple to win a scientific Nobel Prize, which they did in 1903. Marie was born in Poland but eventually moved to France to study at a university, where she met Pierre. Together, they discovered unknown elements that came from the radiation products of uranium salt.

- The first scientist to compile a list of the elements was John Dalton. He listed all twenty elements known at the time of his writing and arranged them by relative mass, having given hydrogen a mass of 1. In the years that followed, scientists continued to find new elements in nature to add to the list. By the end of the 1860s, chemists had found 63 different elements. The modern periodic table is considerably larger, as there are now 118 known elements! Of these 118 elements, only 94 occur naturally on Earth, while the rest are man-made.

- There are a few elements that have properties of both metals and nonmetals. These are called metalloids. The most important of these in our modern world is silicon. Metalloids are semi-conductors, which means that they only conduct electricity (transfer energy between atoms) under some conditions. This property is what allows silicon to be the key ingredient in our computer chips. You can probably guess why silicon is so important considering how much our modern world relies on computers.

- Over 99% of all known compounds are covalently bonded compounds, and most of these are made just of carbon, hydrogen, nitrogen, oxygen, phosphorus, and sulfur.

- Carbon is the master of covalent bonds. This is largely the reason why it is the main building block of living organisms.

- Pure iron reacts with oxygen to make rust. This is why playground equipment made of iron, like the chains of swings, begins to rust after years spent out in the open air (out in the oxygen). It is probably also why playground equipment now tends to be made not of metals like iron but of high-density plastics.

- A bike tire must contain a certain amount of air in order to maintain its shape (its firmness), allowing you to ride your bike. If there is a small hole in the tire, air leaks out, meaning there is less pressure being exerted on the tire. When this happens, it goes flat, and that is the end of your fun day of riding bikes with your friends.

- The air in the atmosphere is constantly pushing down on your skin—we refer to this as atmospheric pressure. If you are at sea level, every square inch of your skin experiences 14.7 pounds of weight from the atmosphere. But as you go up in elevation and the column of air above you becomes smaller, you experience less air mass pushing down on you and the atmospheric pressure decreases. This is why the air is "thinner" in the mountains.

- Gases and liquids take the shape of the container in which they are found.

- Unlike the gaseous phase, the liquid phase is very hard to compress. Imagine trying to squeeze a full water balloon into a smaller space. If you try hard enough, you will most likely end up wet, as the balloon pops due to the force of water pushing out. As there is little leftover room between liquid molecules, it takes a lot of energy to be able to force them to get closer to each other.

*Remember:*

*Dry ice is an example of sublimation.*

- A substance's melting point and freezing point are the same temperature; the difference is that the melting point refers to the transition from solid to liquid and the freezing point is the transition from liquid to solid. This temperature is the same in both cases and is particular to different substances. The state the matter transfers to depends on which direction the energy is moving (are you heating it up or cooling it down?).

- We have looked at the transition of solids to liquids, which is the transition we most often observe. But it is possible to transition directly from a solid to a gas phase (in other words, to skip over the liquid phase). This change is called sublimation. The main example you might be familiar with is the transition of dry ice to a gas. Dry ice is solid carbon dioxide and it must remain very cold. It goes from the solid phase to the gas phase at -78.5 °C.

- The hardest solid we know of is a form of carbon we call diamonds. Diamonds are neither ionic nor molecular solids. They are an example of what is called a network solid. Network solids are made of very large numbers of atoms covalently bonded to each other. Diamonds are large networks of covalently bonded carbon atoms.

# VARIOUS PERIODIC TABLE ELEMENTS

# KEY TERMS

**Absolute quantity** – *Chapter 3*: A measurement that can never be negative, for example speed (the lowest speed is standing still, at zero).

**Acceleration** – *Chapter 3*: The technical term for the change of velocity with time; this means that an object accelerates when its speed changes, when the direction of movement changes, or when they both change.

**Air resistance** – *Chapter 5*: A kind of drag where an object's motion is slowed by the air through which it is moving.

**Alpha particle** – *Chapter 9*: A form of radiation that is heavy and has a positive charge equal in magnitude to two electrons; when collected, alpha particles turn into helium as they collect electrons and become neutral.

**Anion** – *Chapter 10*: An atom that gains electrons and so has more electrons than protons and therefore has a net negative charge.

**Atomic model** – *Chapter 9*: First proposed by Democritus and Leucippus in ancient Greece as an attempt to understand the changes of matter seen in nature as one substance changes to another or as living things grow.

**Atomic number** – *Chapter 10*: The number of protons an element has; elements on the periodic table are arranged by it.

**Axes** – *Chapter 3*: In a Cartesian coordinate system, the horizontal (x axis) and vertical (y axis) lines that allow you to chart data in relation to an origin point.

**Beta particle** – *Chapter 9*: A form of radiation that is negatively charged, with the charge equal to that of one electron; much smaller than an alpha particle.

**Biological sciences** – *Chapter 1*: Areas of science that study living things.

*Remember:*
*Over 99% of all known compounds are covalently bonded compounds.*

**Boiling temperature** – *Chapter 12*: The temperature at which liquid changes to gas.

**Cation** – *Chapter 10*: An atom that has lost one or more electrons, giving it more protons than electrons and therefore a net positive charge.

**Chemical bond** – *Chapter 10*: The force of attraction between two distinct atoms to make a new entity.

**Combustion reaction** – *Chapter 11*: A chemical reaction in which a "fuel" is oxidized with oxygen gas.

**Compound** – *Chapter 1*: A chemical substance made up of multiple elements, and the elements can be broken up or joined through a chemical reaction.

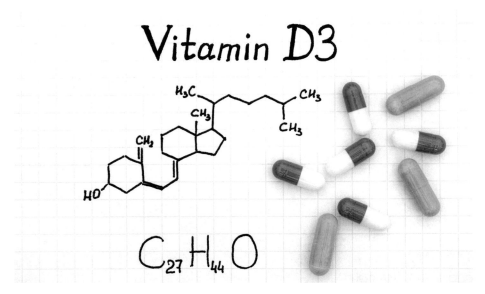

**Compression** – *Chapter 12*: The decrease in volume of a substance from added stress applied to it—gases are the only state of matter that can be compressed a significant amount.

**Condensation** – *Chapter 12*: When energy is removed from a gas through cooling it to turn it into a liquid.

**Conductors** – *Chapters 6 and 8*: Materials that readily get hot or cold and are used when heat needs to be moved; also a term given to things that transfer electrical currents easily.

**Coulomb's Law** – *Chapter 8*: Developed by Charles-Augustin de Coulomb, a law in chemistry that states the force of attraction or repulsion between charges is proportional to the product of the charges (so the stronger the charge, the stronger the attraction or repulsion), and the force gets weaker the farther apart the charges are from each other; the unit of measurement for charges is called the Coulomb.

**Covalent bond** – *Chapter 10*: A bond in which, instead of gaining or losing electrons, the atoms "share" the electrons between them to form a bond that keeps them together; over 99% of all known compounds are covalently bonded compounds, and most of these are made just of carbon, hydrogen, nitrogen, oxygen, phosphorus, and sulfur.

**Density** – *Chapter 1*: Describes a relationship between volume and mass—it is the amount of mass packed into a given unit of volume; it can be used as a descriptor of how heavy an object is.

**Drag** – *Chapter 5*: The slowing of an object's motion caused by the substance through which the object is moving; can occur through *air resistance* and *hydrodynamic drag*.

**Electric circuit** – *Chapter 8*: A closed loop made of wires attached to both sides of a voltaic cell for charged particles to move through; found in essentially all modern technology (refrigerators, cell phones, etc.).

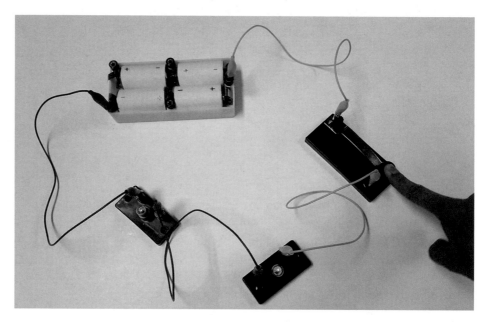

**Electric potential** – *Chapter 8*: The push behind an electrical current.

**Electrical current** – *Chapter 8*: Energy made of moving charges (charged particles) used to do some useful work, like provide power to a television.

**Electricity** – *Chapter 8*: A form of energy generated from negatively and positively (opposite) charged atoms reacting to each other.

**Electromagnetic waves** – *Chapter 7*: Pure transfers of energy that require no matter to move the energy (though they can move through matter as well).

**Electromagnetism** – *Chapter 8*: The science of charge and the forces and fields associated with it—consists primarily of the study of electricity and magnetism.

**Electron** – *Chapter 9*: A negatively charged subatomic particle in an atom; its mass is so much smaller than the other two subatomic particles (protons and neutrons) as to be considered negligible.

**Electron cloud** – *Chapter 9*: The area of an atom outside its nucleus, where all the electrons are found.

**Element** – *Chapters 1 and 10*: A pure substance made of the same atom that cannot be broken into any other type of substance by using chemical reactions, where the type is given by the number of protons in the atom's nucleus.

**Endothermic reaction** – *Chapter 11*: A chemical reaction that requires that energy be added to it for the reaction to occur.

**Energy** – *Chapter 6*: The capacity ("ability") to do work.

**Entropy** – *Chapters 6 and 10*: A measure of unavailable (lost) or wasted energy (heat). Also a measure of molecular disorder.

**Exothermic reaction** – *Chapter 11*: A chemical reaction that gives off energy (most often in the form of heat).

**First law of motion** – *Chapter 4*: Developed by Sir Isaac Newton, this law states that an object in motion will continue in motion and an object at rest will continue at rest unless an outside force acts upon it.

**Frequency (wave)** – *Chapter 7*: How many waves are measured in a specific point in space for a given unit of time.

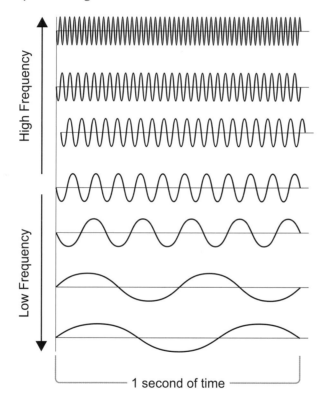

**Friction** – *Chapter 5*: A force that acts on an object in the opposite direction of the object's movement; it is the result of contact between two solid surfaces.

**Force** – *Chapter 4*: Any pull or push that changes an object's motion.

**Function** – *Chapter 2*: A mathematical relation between two or more variables.

**Gamma rays** – *Chapter 9*: Electromagnetic waves of very high frequency.

**Gravity** – *Chapter 5*: An attraction (an invisible force) that pulls together all objects made of matter.

**Heat capacity** – *Chapter 6*: How easily some mass of a substance changes temperature—how quickly it heats up or cools down.

**Heterogeneous mixture** – *Chapter 1*: A mixture in which we can observe its different components.

**Homogeneous mixture** – *Chapter 1*: A mixture in which the different substances that are mixed together *cannot* be observed separately.

**Hydrodynamic drag** – *Chapter 5*: A kind of drag where an object's motion is slowed by the fluid through which it is moving.

**Hypothesis** – *Chapter 2*: A proposed explanation for an observed phenomenon.

**Impulse** – *Chapter 4*: A force that acts upon an object for a short period of time.

**Inertia** – *Chapter 4*: A property of matter that describes how an object in motion or at rest will continue in that state unless an outside force acts upon it to stop it or move it; part of Newton's first law of motion.

**Instantaneous speed** – *Chapter 3*: The speed of an object at any particular moment in time.

**Insulator** – *Chapters 6 and 8*: Material that does not easily change temperature and is used when we need to keep heat in or out of somewhere; also a term used to describe materials that prevent the flow of electrical currents.

**Ionic bond** – *Chapter 10*: The attraction of ions to one another due to their opposite charges.

**Ionic compound** – *Chapter 10*: A compound formed from ionic bonds.

**Ions** – *Chapter 10*: Charged atoms; can be either cations (positively charged) or anions (negatively charged).

**Isotopes** – *Chapter 9*: Atoms of the same element that have different masses.

**Kinematics** – *Chapter 3*: The study of motion.

**Kinetic energy** – *Chapter 6*: The energy of objects that are in motion and/or doing work.

**Law of conservation of energy** – *Chapter 6*: A proven explanation of how energy is neither created nor destroyed, only converted or transferred from one kind of energy to another.

**Law of conservation of mass** – *Chapter 9*: Proposed by Antoine Lavoisier, this law states that matter can neither be created nor destroyed.

**Law of definite proportions** – *Chapter 9*: A proven explanation of how all chemical compounds are made of whole number mass ratios of the elements in that compound, and that they are fixed and constant.

**Law of multiple proportions** – *Chapter 9*: A proven explanation of how two different compounds that are made of the same elements have different whole number mass ratios.

**Law of nature** – *Chapter 2*: A stated scientific fact that cannot change with time—it has been proven to be constant (e.g., the law of gravity); it gives a mathematical explanation to observed phenomena.

**Liquefaction** – *Chapter 12*: The process of a substance turning from a solid into a liquid via heat being applied to it.

**Lodestones** – *Chapter 8*: Stones once thought to be magical that attracted ferrous metals to them; used in early navigation as the first magnetic compasses, stemming from a Middle English word that meant "leading stone."

**Longitudinal wave** – Chapter 7: A wave that travels in the same direction as the disturbance causing it (they travel parallel to each other).

**Magnetism** – *Chapter 8*: A phenomenon produced by the motion of an electric charge, resulting in an attraction (things coming together) or a repulsion (things pushing apart).

**Mass** – *Chapter 1*: The quantity of matter in a given object.

**Mass ratio** – *Chapter 9*: The balance of the various elements' masses in a chemical compound.

**Matter** – *Chapter 1*: Anything that takes up space and can be weighed, which means it has volume and mass, or density.

**Mechanical waves** – *Chapter 7*: Waves that transfer energy by moving matter (as they travel through matter) in an oscillating or repeating pattern.

**Mechanics** – *Chapter 3*: The study of how we understand the movement of objects in space.

**Metalloids** – *Chapter 10*: Elements that have properties of both metals and nonmetals.

**Mixture** – *Chapter 1*: Something that is made from at least two or more pure substances; there are two types of mixtures: heterogeneous mixtures and homogeneous mixtures.

**Molecular elements** – *Chapter 10*: Atoms of the same element bonded to each other.

**Molecules** – *Chapter 10*: Atoms that are covalently bonded to one another.

**Momentum** – *Chapter 3*: A measure of movement that calculates the total motion of an object by multiplying its mass by its velocity.

**Motion** – *Chapter 3*: The movement of objects through space.

**Neutron** – *Chapter 9*: A subatomic particle found in the nucleus of an atom (along with the protons) that has no charge, meaning it is neutral; it is a tiny bit bigger than a proton.

**Nucleus** – *Chapter 9*: The core of an atom.

**Oxidation reaction** – *Chapter 11*: A chemical reaction where oxygen is added to an element, molecule, or compound.

**Parabola** – *Chapter 5*: The arc shape an object makes when it is hurled in the air and brought back to the ground by the force of gravity.

**Periodic table** – *Chapter 10*: A chart of chemical elements arranged by their atomic number and grouped by similar chemical characteristics.

**Physical change** – *Chapter 1*: A change in a given substance that does not change the *kind* of substance (ice changing to water, for example).

**Poles** – *Chapter 8*: The two ends (north and south) of a magnet.

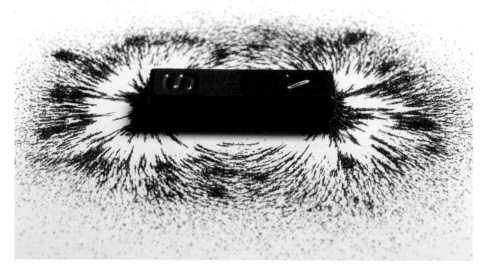

**Potential energy** – *Chapter 6*: Energy that is stored up to be able to do work.

**Pressure** – *Chapter 12*: The ratio of force divided by area.

**Product** – *Chapter 11*: A substance that is formed as a result of a chemical reaction.

**Properties** – *Chapter 1*: Descriptors or characteristics of a certain substance (such as color, texture, density, etc.).

**Proton** – *Chapter 9*: A positively charged subatomic particle in an atom, equal in magnitude to the negative charge of an electron, meaning that one proton will perfectly neutralize the charge of one electron.

**Pure substances** – *Chapter 1*: Substances that cannot be separated from others using physical changes; there are two types of pure substances: *elements* and *compounds*.

**Qualitative measurement** – *Chapter 2*: A measurement that uses words to describe a given substance or object, such as its color or the way it feels or smells.

**Quantitative measurement** – *Chapter 2*: A measurement that uses a fixed number (a quantity) to describe a given substance or object, such as how many pounds it weighs or how tall something is in inches.

**Radiation** – *Chapter 9*: Energy put out from a source in the form of waves or particles.

**Reactants** – *Chapter 11*: Also called reagents, these are the initial chemicals that are grouped and arranged to form a chemical reaction.

**Reduction reaction** – *Chapter 11*: A chemical reaction where oxygen is removed from an element, molecule, or compound.

**Reflection** (wave) – *Chapter 7*: The bouncing back of a wave, as with the echo of a sound wave.

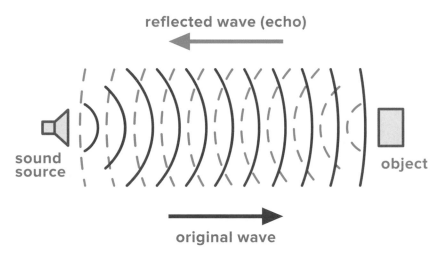

reflected wave (echo)

sound source

object

original wave

**Refraction** – *Chapter 7*: The bending or redirection of light as it passes from one medium to another.

**Resonance** – *Chapter 7*: The louder sound effect when waves of the same frequency vibrate with each other and their effects get amplified, or added to one another.

**Scientific method** – *Chapter 2*: The method science uses to study the material world, or matter; consists of several steps that are generally listed as: (1) observation, (2) measurement/research, (3) experiment, (4) hypothesis formulation, (5) testing/analyzing of the hypothesis, (6) modification of hypothesis (if necessary).

**Second law of motion** – *Chapter 4*: Developed by Sir Isaac Newton, the proven explanation that states that the acceleration of an object of constant mass is proportional to the force applied on the object.

**Seismic wave** – *Chapter 7*: A wave in the Earth caused by an earthquake.

**SI System (International System)** – *Chapter 2*: An international system of measurement that uses physical units; it provides a defined and agreed upon unit for each of the basic measurements: mass, length, time, volume, etc.

**Solidification** – *Chapter 12*: The process of a substance turning from a liquid into a solid when it cools.

**Speed** – *Chapter 3*: A measure of how fast an object is moving, calculated by dividing the length of the object's motion (inches, meters, yards, miles, etc.) by the time it took for the object to cover that distance (seconds, minutes, hours).

**Static electricity** – *Chapter 8*: An electric charge that is not moving (static means stationary) but rather held in some object, like in your finger after you drag your feet on the carpet.

**Strong force** – *Chapter 9:* A force that counteracts or works against the electric repulsion of the positive charges of the protons.

**Sublimation** – *Chapter 12*: The change from a solid state straight to a gaseous state, skipping over the liquid state.

**Theory** – *Chapter 2*: A systemic explanation of a natural phenomenon that can predict the results to be observed in a situation.

**Thermal equilibrium** – *Chapter 6*: When an object's temperature and the environment in which it is found have the same temperature; they are in a *state of balance* in regard to their temperature.

**Third law of motion** – *Chapter 4*: Developed by Sir Isaac Newton and known as the action-reaction law, this proven explanation states that when a force is applied to an object, the object applies a *reaction force* equal in size to the force applied to it—but in the opposite direction.

**Transverse wave** – *Chapter 7*: A wave that moves at a right angle, or perpendicular, to the disturbance causing it.

**Vacuum** – *Chapter 7*: A region of space with little or no matter.

**Vaporization** – *Chapter 12*: When energy (heat) is added to a liquid and it turns into a gas.

**Velocity** – *Chapter 3*: A measurement that describes the speed of an object while also factoring in the direction of its motion.

**Volt** – *Chapter 8*: The unit used to express electrical potential; named after Alessandro Volta, an Italian scientist.

**Volume** – *Chapter 1*: A measurement that refers to the amount of space something occupies.

**Wave (sound & light)** – *Chapter 7*: Transfer of energy created by a disturbance.

**Wavelength** – *Chapter 7*: The distance between repeated parts of the oscillating pattern of a wave.

**Wave speed** – *Chapter 7*: How much distance the wave covers in a given unit of time (seconds, minutes, hours).

**Weight** – *Chapter 5*: A measurement of the force of gravity with which an object is attracted toward another object.

**Work** – *Chapter 6*: The transfer of energy that takes place when a force acts within a certain distance.

# IMAGE CREDITS

p68-69  Organ of the St. Louis Cathedral, Notre-Dame in Poissy, Paris, France © Stas Knop, Shutterstock.com

p70-71  An induction coil or spark coil, (previously known as an inductorium or Ruhmkorff coil) is a type of electrical transformer used to produce high-voltage pulses from a low-voltage direct current supply © Andrey Semenov, Shutterstock.com

p72  Power lines © SSSCCC, Shutterstock.com

p73  Discovering static electricity with with hair and an inflated balloon © Raylui321t, Shutterstock.com

p73  Static electricity illustration © tonkhao wanpiya, Shutterstock.com

p74  Cross section of high-voltage cable © Pedal to the Stock, Shutterstock.com

p75  Luigi Galvani discovered that frog's legs twitch when electricity is passed through the muscles, a phenomenon called galvanism that lead to the subject of electrophysiology and treatment by electrotherapy. Source: David Ames Wells, *The science of common things: a familiar explanation of the first principles of physical science. For schools, families, and young students.* Publisher Ivison, Phinney, Blakeman, 1859, 323 pages. Author: Luigi Galvani. [public domain], via Wikimedia Commons

p75  Leyden battery. Source: https://www.flickr.com/photos/adamcoop68/10063726865. Author: Flickr; Adam Cooperstein. License: (CC BY 2.0), https://creativecommons.org/licenses/by/2.0/deed.en

p75  Magnetite rock attracting paper clips © Michael LaMonica, Shutterstock.com

p76  Magnetic field illustration © udaix, Shutterstock.com

p76  Faraday's first Experiment (engraving) / Illustration for Cyclopaedic Science Simplified by J H Pepper (Frederick Warne, 1869). / English School, (19th century) / English / Photo credit © Look and Learn / Bridgeman Images

p77  Faraday's experiment to try to induce a current from a magnetic field, with a battery on the left, an iron ring in the centre, and a galvanometer on the right. This diagram is based on one found in page 263 of Physics: Principles with Applications, fifth edition, author Douglas C. Giancoli, illustrators Patrice Van Acker and Tamara Newnam Cavallo. Source/Author: Eviatar Bach. License: (CC0 1.0), Public Domain Dedication. https://creativecommons.org/publicdomain/zero/1.0/deed.en

p77  Electric copper coil inductor © PitukTV, Shutterstock.com

p78  Benjamin Franklin, philosopher, physicist and American ambassador, experimenting with his lightning rod mounted on a kite. September 1752 / Reid, Stephen (1873-1948) / English / Photo credit © Giancarlo Costa / Bridgeman Images

p79  Portrait of Benjamin Franklin (colour litho) / Duplessis, Joseph Siffred (1725-1802) (after) / French / Photo credit: Peter Newark American Pictures / Bridgeman Images

p80-81  3D atom model with yellow and blue particles. The central nucleus are surrounded by a cloud of negatively charged electronse © Dabarti CGI, Shutterstock.com

p82  Close up view of iron ore © Frau aus UA, Shutterstock.com

p82  Molten metal casting in foundry. Filling mold with hot liquid iron and producing iron components in steel plant © Aleksandar Malivuk, Shutterstock.com

p82  Screwdrivers © David Fadul, Shutterstock.com

p83  Timeline of atomic models © Designua, Shutterstock.com

p84  Law of conservation of mass and physical change. Ex. ice cubes and water © Designua, Shutterstock.com

p85  3D cubes © Vector FX, Shutterstock.com

p86  Portrait of John Dalton (1766-1844), English physicist and chemist / Photo credit Stefano Bianchetti / Bridgeman Images

p87  Atomic & Nuclear Research, 1897-1939 J J Thomson's cathode ray tube with magnet coils, 1897. This apparatus was used to dicscover the electron. In 1896, in Cambridge, Joseph John Thomson (1856-1940) began experiments on cathode rays. In Britain, physicists argued these rays were particles, but German physicists disagreed, thinking they were a type of electromagnetic radiation. Thomson showed that the cathode rays were particles with a negative charge and much smaller than an atom. He published this information in April 1897; the particles were later named electrons. Apparatus shown without its stand. ©SSPL/Science Museum / Photo credit: SSPL/UIG / Bridgeman Images

p87  JJ Thomson atomic model diagram © Dream01, Shutterstock.com

p87  Plum Pudding. Source: https://www.flickr.com/photos/dennissylvesterhurd/343148974. Author: Dennis Sylvester Hurd. License: (CC0 1.0), Public Domain Dedication. https://creativecommons.org/publicdomain/zero/1.0/deed.en

p88  Radiation penetration types. Permeability power, alpha, beta, x ray, gamma rays. Neutron particle. paper sheet, wood, iron plate, water © grayjay, Shutterstock.com

p89  Pierre (1859-1906) and Marie Curie (1867-1934) in their laboratory, c.1900 / Gribayedoff, Valerian (1858-1908) / Russian / Archives Larousse, Paris, France / Photo credit: Bridgeman Images

p89  Atomic structure: Proton, Electron and Neutron. © KenAge, Shutterstock.com

p90  The basic model of the Oxygen atom containing protons, neutrons and electrons © bsr00, Shutterstock.com